Energy Scenarios for the Future

Cornel Stan

Energy Scenarios for the Future

Ways to Climate-friendly Mobility, Heating and Industry

 Springer

Professor Dr.-Ing. habil. Prof. E.h. Dr. h.c. mult. Cornel Stan
FTZ – Research and Technology Association
West Saxon University
Zwickau, Germany

ISBN 978-3-662-69686-6   ISBN 978-3-662-69687-3   (eBook)
https://doi.org/10.1007/978-3-662-69687-3

Translation from the German language edition: "Klimagerechte Energieszenarien der Zukunft" by Cornel Stan, © Springer-Verlag GmbH Deutschland, part of Springer Nature 2024. Published by Springer Berlin Heidelberg. All Rights Reserved.

This Springer imprint is published by the registered company Springer-Verlag GmbH, DE, part of Springer Nature.
The registered company address is: Heidelberger Platz 3, 14197 Berlin, Germany

If disposing of this product, please recycle the paper.

Preface

Not combustion engines and other fireplaces need to be replaced, but what they have to burn.

"To limit global warming to 1.5°C, greenhouse gas emissions must peak by 2025 at the latest and fall by 43% by 2030," said the United Nations Intergovernmental Panel on Climate Change. 2030, there are very, very few years until then. *Decarbonization* means reducing carbon by 43% in six to seven years. Simply put, *we should eliminate the burning of fossil fuels in order to save the planet's climate.*

Eliminate fossil fuels, when their consumption is so increasing?

Does anyone seriously believe that Russia will give up more than half of its state budget from natural gas and oil exports? Should Saudi Arabia give up almost half of its gross domestic product, which is generated by oil sales? Or should Kuwait give up almost all of its export earnings that come from oil?

The U.S., Saudi Arabia and Russia produce 42% of the world's oil.

Natural gas and petroleum, the classic representatives of fossil fuels, are heavily consumed in Asia (China, Japan, Korea) to produce electricity for the millions of zero-emission electric cars they build and sell. 63% of the world's consumption of oil and natural gas in *China,* Japan and Korea is quite a lot, although only 14% is produced in these areas.

And it goes on like this: Should *Exxon Mobil, Chevron, Shell, Total,* switch from oil and gas to photovoltaics and wind turbines immediately?

The main counter-argument is the Paris Agreement, which was agreed by 195 countries during the UN Climate Change Conference (COP21) on 12 December 2015. Among its three main objectives is also:

„Limiting the increase in the global average temperature of the Earth's atmosphere to 1.5 degrees Celsius if possible, but in any case, well below two degrees Celsius compared to the pre-industrial era."

However, the Europeans then decided on their own, even more ambitious climate protection package:

"Fit for 55"?

Profiting on this occasion, the EU states also agreed on a reform of the *EU Emissions Trading System (ETS),* for the establishment of a social climate fund worth more than 80 billion euros, for a new *CO2 border adjustment* system (CBAM), for a separate emissions trading system for transport and buildings, and for new rules **for emissions trading** in aviation and shipping.

Fit for 55 then the EU Emissions Trading System decided:

"The EU's CO2 emissions must be reduced by 55 **percent** by 2030 compared to 1990. *(what does that have to do with the year 1990?).* By 2050, Europe is to become **greenhouse gas-neutral**, i.e. practically "decarbonized"!

And so, politics and far too many entertaining media of all kinds are currently dealing with *the global economy, prosperity principles such as home heating and mobility* every day. It is precisely these issues that are usually radically sacrificed on the altar of saving the climate. Ignoring physics, thermodynamics, and technical diversity, monopoly-like, universal solutions are decreed: automobile drives are only electric. In addition, heat pumps with heat from the cold ambient air, so, as a thermodynamics scholar, I may be allowed to call them *backward-running, air-sucking "apartment refrigerators"*. As a heating open, of course, electricity for the large and small, very diverse industry only from wind turbines and solar panels, although their contribution has remained almost negligible for decades. On the basis of comprehensible connections between physics and thermodynamics, this book represents an attempt to demystify such bitter and dangerous climate fairy tales and to build efficient, real scenarios against them.

Wind and sun are certainly good and right, but far from the real necessity.

Plant residues, algae, waste oils and fats, it describes "green" hydrogen as a storage or intermediate storage medium. But it also describes surprising, unexpected **energy scenarios** that arise from the interlinking of otherwise common thermal machines: a jet engine, combined with a steam power plant, the *diesel engine* as a **combined heat and power plant.**

The book describes technical solutions that are more efficient than pure electric cars and air-based home heat pumps. It compares photovoltaics and wind energy with new, climate-friendly energy sources and describes process chains.

In the first chapters of the book, the engineering and physical fundamentals of **photovoltaic systems** and **wind turbines** are presented and documented with concrete examples in comparison to other power plants, hydroelectric power plants and nuclear power plants. Whether new solutions and methods, such as on-shore and off-shore wind turbine parks and photovoltaic panels on roofs and balconies, in the desert and on water, are efficient solutions, will be exemplified and commented on.

One chapter is devoted to **electric cars**: promising *"combustion engines" with green hydrogen, hydrogenated frying fats, ethanol from sugar cane, with concrete examples and results are presented.*

In a further chapter, the **heat pump** with air, which is much discussed in the media and in politics, i.e. the *"refrigerator with a reverse cycle of the working fluid that uses cold ambient air as a heat supply"*, will be equipped with usable solutions according to thermodynamic and engineering criteria, juxtaposed with each other.

A separate chapter is dedicated to **climate-friendly** fuels on the resources, potentials and properties. Methanol as e-fuel, ethanol from fry fats, hydrogenated vegetable oil residues (HVO-hydrogenated vegetable oils), processed plaster residues and animal fats are considered individually.

At the beginning of the respective chapters a very short and compact "**executive summary**" is appended. A preamble of a few lines, often a quote, first describes the real problem. The reader may also be surprised at times. The rest, is a self-commentary, often sharp, by the author of this book, with analysis and solution of the problem, obtained on the basis of his many years of experience in research projects with global companies, with teaching in many universities from San Francisco and Paris to Turin or Brasov.

The work is of interest not only to students of technical and economic sciences who have a background in physics and technical thermodynamics, but also to executive people in politics, business and the media.

Cornel Stan   Zwickau, Germany, June 2024

Table of contents

Chapter 1

Current Energy Scenarios: Decarbonization

*"To limit global warming to 1.5°C, greenhouse gas emissions must peak by 2025 at the latest and fall by 43% by 2030," said the United Nations Intergovernmental Panel on Climate Change. 2030 is seven years from now. Decarbonization, means reducing **carbon** by 43% in seven years. However, carbon is not only the basis of our DNA, but also that of all hydrocarbons such as crude oil (heptane, octane, pentane) or natural gas (methane) or pure coal. The answer may be simple: eliminate the burning of fossil fuels in order to save the planet's climate.*

Does anyone seriously believe that Russia will give up more than half of its state budget from natural gas and oil exports?

Should Saudi Arabia give up 42 percent of its gross domestic product, which is generated by oil sales?

Or should Kuwait give up 90 percent of its export earnings?

The <u>United States, Saudi Arabia and Russia produce almost half (42%)</u> of the world's oil.

It is also interesting in which regions of the world this oil is **consumed**: Asia (China, Japan, Korea) consume 63%, although in these countries only 14% are produced. The Middle East, on the other hand, consumes only about 10% of its petroleum products, which is 53%.

The largest **natural gas exporter** in the world is Russia with a share of 19.1% (2019), followed by Qatar and the USA with around 10% each.

In terms of **coal production and consumption**, Asia and the Pacific make up an overwhelming part of the world.

Africa has so far accounted for a very moderate share of coal consumption. But what happens when China builds so many coal-fired powerplants in Africa? **The Chinese, of course, will also bring the coal.**

Who benefits from the many organizations around the world that talk and study about it when the world is burning?

IPCC (Intergovernmental Panel on Climate Change), **UNFCCC - United Nations Framework Convention on Climate Change**, UNEP - United **Nations Environment Programme, WWF - World Wide Fund for Nature**

And there is an annual report: just the core team of the assessment report included 721 experts from 90 countries, and nearly 3,000 people from all over the world were sent by governments and organizations to participate in its preparation.

The 40-page summary, written specifically for the politics of the world, does not contain any concrete causes, actions or solutions, but only insights into what has changed or changed in the Earth's climate.

Instead of turning model countries into model regions of greenhouse gas neutrality, these numerous organizations should urgently rescue the many poor countries in Africa, Asia and South America from oil, gas and coal, not through money, but through the concrete construction of high-tech plants that produce regenerative, climate-neutral or climate-friendly forms of energy! Centralized, decentralized, simple, complex. Every vote counts.

1 Current Energy Scenarios: Decarbonization

1.1 The Paris Agreement, Fit for 55, Processes Greenhouse Gas Neutral

> *Everything must be made as simple as possible. But not simpler. Makes everything as easy as possible. But not simpler (Albert Einstein, 1879 – 1955).*

Do we want to banish "*carbon*" from our universe "*never to be seen again*"? It would be fatal: carbon and its chemical compounds in molecules of various kinds are the basis of our existence, our life par excellence, starting with the DNA (deoxyribonucleic acid) of all living beings.

Let's leave an unfortunate American formulation (simply: "carbon", rather than a looser form for <u>carbon dioxide</u>). The carbon should only be "neutralized":

<u>The Paris Agreement </u>was signed during the UN Climate Change Conference (COP21) on 12 December. December 2015 by 195 countries. Among its three main objectives is also:

"Limiting the increase in the <u>global average temperature of the Earth's atmosphere to 1.5 degrees Celsius if possible</u>, but in any case, well below two degrees Celsius compared to the pre-industrial era."

The United Nations Intergovernmental Panel on Climate Change had previously pointed out that exceeding the 1.5-degree threshold risks triggering far more severe climate change impacts, including more frequent and *severe droughts, heat waves and rainfall.*

And another quote: "To limit global warming to 1.5°C, greenhouse gas emissions must peak by 2025 at the latest and fall by 43% by 2030".

The **Paris Agreement is in itself a milestone** in the multifaceted climate change process, because for the first time a binding agreement brings all nations together to combat climate change and adapt to its impacts.

However, the Europeans then decided on their own, even more ambitious climate protection package:

„Fit for 55". What is that?

On this occasion, the EU states also agreed on a reform of the *EU Emissions Trading System (ETS),* for the establishment of a social climate fund worth more than 80 billion euros, for a new *CO_2 border adjustment* system (CBAM), for a separate emissions trading system for transport and buildings, and for new rules for emissions trading in aviation and shipping.

Fit for 55 then decided: "The EU's CO_2 emissions must fall by 55 percent by 2030 compared to 1990 (hence fit for 55 – *but what does that have to do with the year 1990?*). By 2050, Europe is to become **<u>greenhouse gas-neutral</u>**, i.e. practically "decarbonized".

UN Secretary-General António Guterres recently received a new edition of the *Intergovernmental Panel on Climate* Change **(IPCC) *assessment report***, and his reaction after first reading the 40-page summary conclusions was: "*I have read many reports in my life, but never anything like this... The world can hardly be saved.*"

The core team of the assessment report included 721 experts from 90 countries, and nearly 3,000 people from around the world were sent by governments and organizations to participate in the drafting.

The report of Working Group Number I (of III) comprises 3949 pages and was prepared by 234 scientists from 66 countries. The report analyzes and synthesizes 14,000 scientific studies from various disciplines. The 40-page summary, written specifically for the politics of the world, does not contain any concrete *causes, actions or solutions,* but only *insights* into what has changed or changed in the Earth's climate.

So, if they still don't know exactly what to do, policymakers prefer to ignore such an avalanche of sophisticated and scientifically and legally rehashed data.

However, there are numerous committees, panels and organizations from the world to the local level:

UNFCCC - United Nations Framework Convention on Climate Change, which deals with the prevention of anthropogenic disturbances and measures to slow global warming. The UNFCCC Secretariat in Bonn alone employs 450 people!

UNEP - United Nations Environment Program based in the Kenyan capital Nairobi with 800 staff.

WWF - World Wide Fund for Nature. It was founded in 1961 as one of the largest international nature and environmental protection organizations. The company employs 6,200 people and generates an annual turnover of almost 700 million euros.

In addition, internationally, but also in almost all countries, there are hundreds, if not thousands, of environmental organizations with similar goals.

What are all these organizations, all these people doing that have huge resources at their disposal to counteract the exploding carbon dioxide emissions into the atmosphere?

The answer may be simple: eliminate the burning of fossil fuels in order to save the planet's climate and gain energy independence from their exporters.

The relationship between carbon dioxide and atmospheric temperature is presented in various forms, more or less complex, more or less scientific, often very much from the perspective of the said author.

A short, simplified explanation by the author of this book, for a basic understanding of the connections, seems necessary at this point:

The average temperature of the Earth's atmosphere of about 15 °C is a consequence of the **natural** greenhouse effect, which is caused by multiatomic gas such as water vapors (62.4%), carbon dioxide (21.8%), ozone (7.3%), methane and nitrous oxide (8.5%) as shown in Figure 1.1. However, the carbon dioxide already present in the atmosphere is added annually at a rate of 0.6% caused by human activities, namely the burning of fossil fuels (*coal, diesel, gasoline, natural gas*). This cumulative annual excess in the atmosphere increases the participation of carbon dioxide in these proportions.

Studies show [1], [2] that without the natural greenhouse effect produced by these gases (i.e. without the excess carbon dioxide produced over time), the average temperature of the Earth's atmosphere would drop by 33 °C, equivalent to minus 18 °C.

For the sake of clarity, the natural greenhouse effect can also be explained in a simplified form: the *sun's rays* move towards our planet with great *intensity* and high *frequency*. These two components (*intensity and frequency -or, as the reciprocal of frequency, the wavelength*) form the *energy flow* of each beam. A ray of sunlight that penetrates the atmosphere reaches the Earth and emits part of it as a *stream of heat* to the earth's crust, fauna, flora and water of the seas and oceans. The heat *flow* provides an energy output (kilowatts), which is converted into the beneficial *heat* (kilowatt hour) over time. The sun's ray is then reflected towards the atmosphere, but its intensity and frequency become less compared with the portion of the *energy* that has been emitted to the Earth. Caused by polyatomic gases in the atmosphere, such as, for example, carbon dioxide, which is similar in structure to silicon dioxide (i.e., glass that usually covers classic greenhouses), it cannot fully penetrate this gaseous "roof". Such radiation is then partially deflected back to Earth. And this back and forth repeats itself many times. The sun's ray, even if it becomes weaker and weaker in intensity, always leaves a part of warmth to the fauna, the flora, the earth's crust and the waters.

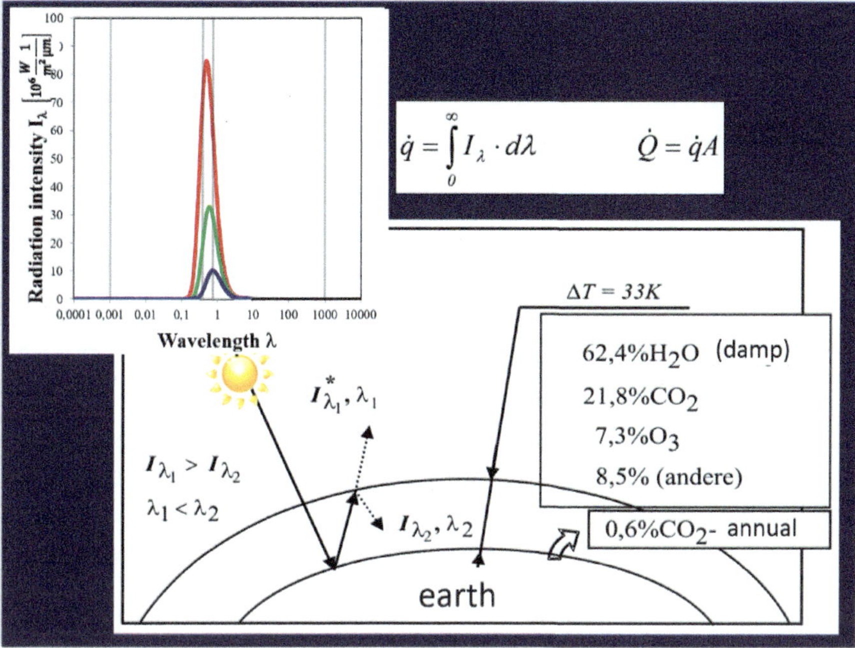

Figure 1.1 Radiation spectrum of incident and reflected electromagnetic radiation

Since the beginning of industrialization, after the invention and introduction of the steam engine (1712), the earth's atmosphere has warmed by an additional $1\,°C$ in this way. It is worth noting that at the same time, the concentration of carbon dioxide in the Earth's atmosphere has increased from 280 [ppm] (*parts per million /parts of air*) to 417 [ppm] (*measured in July 2021 at Mauna Loa Station, Hawaii*). Climate scientists at the *Intergovernmental Panel on Climate* Change (IPCC) have deduced on the basis of corresponding studies [3],[4] that this increase in the amount of anthropogenically produced carbon dioxide in the Earth's atmosphere is responsible for the temperature increase, at least in the last 5-6 decades.

At the current rate of carbon dioxide emissions, the atmospheric temperature increase could reach 5.8°C by the end of this century. So, we have a duty to act quickly: the current goal of the international community is to limit the warming of the Earth's atmosphere to 1.5°C by the end of the century by drastically reducing man-made carbon dioxide emissions.

"Media experts" cheerfully predict: *The excess carbon dioxide is to be absorbed by the many forests and plants of the earth anyway!"* On the other hand, the scientists do not agree:

In 2020, total global carbon dioxide emissions amounted to 34.8 billion tons [5]. In 1960, emissions amounted to "only" 9.3 billion tons. The burning of fossil fuels has therefore undeniably polluted the atmosphere, with

42% is accounted for by electricity and heat generation, 25% by transport and 19% by industry.

It is true that some of the anthropogenic emissions are being reabsorbed by the Earth's oceans and vegetation. Between 1994 and 2007, the Earth's oceans absorbed no less than 34 billion tons of anthropogenic carbon dioxide from the atmosphere, which corresponds to 31% of anthropogenic CO_2 emissions during this period [6]. The main negative effect of this absorption is an increased acidity of the water, which affects both marine flora and marine fauna.

The world's forests absorb 0.55 billion tons annually, making them the largest terrestrial sink. This amount is not immediately "processed" by photosynthesis. The capacity to store carbon dioxide in trees and in the soil beneath them is huge, but it has already reached saturation: 295 billion tons of active biomass, 300 billion tons of organic matter from the soil under trees, 68 billion tons of "dead" wood [7].

A distinction might be helpful: there are trees and trees, there are plants and plants. Maize and millet cannot grow faster due to the increased concentration of carbon dioxide in the air. In contrast, soybeans and wheat grow faster. In tropical forests, lianas grow much faster than other plants due to the increase in carbon dioxide levels in the atmosphere. But side effects must also be considered: lianas prevent the growth of trees, which are good absorbers of carbon dioxide.

One conclusion is unavoidable at this point, with reference to the ambitious "Fit for 55" program: the EU, as an area, is small compared to the rest of the world, greenhouse gas emissions are very low, to insignificant (*less than 10% of total emissions in the whole world*). However, the EU is very developed industrially.

Instead of making Europe a model region of greenhouse gas neutrality, the EU should urgently rescue numerous poor countries in Africa, Asia and South America from oil, gas and coal, not through money, but through the concrete construction of high-tech plants that produce regenerative, climate-neutral or climate-friendly forms of energy!

Easy to say but hard to implement: our planet's climate is undergoing a dizzying process of destruction, but that doesn't seem to interest any fossil fuel seller or buyer, be they states, corporations or private consumers.

Unfortunately, the climate is like viruses: what is not seen and does not hurt does not exist!

Does anyone seriously believe that Russia will give up more than half of its state budget from natural gas and oil exports?

Should Saudi Arabia give up 42 percent of its gross domestic product, which is generated by oil sales?

Or should Kuwait give up 90 percent of its export earnings?

A look at the list of the largest **oil producers** in the world can be very revealing:

The *U.S.* tops the ranking with 19%, followed by *Saudi Arabia* (12%) and *Russia* (11%), followed by Canada, China, Iraq, the *United Arab Emirates, Brazil, Iran, and Kuwait* with around 5% each. The *U.S., Saudi Arabia and Russia produce nearly half (42%)* of the world's oil. And how these states "participate" in international climate conventions can be seen daily in the media!

It is also interesting in which regions of the world this oil is **consumed**: Asia (this time clearly China, Japan, Korea) consume 63%, although in these areas only 14% are produced. The Middle East, on the other hand, consumes only about 10% of its petroleum products, which is 53%.

In America, there is a clear parity between *production* and *consumption*, even if at first glance this is only a balance: America *exports* a lot of oil, but also imports a lot, the reasons for these transactions are purely financial.

The **world's largest natural gas exporter** is Russia with a share of 19.1% (2019), followed by Qatar and the USA with around 10% each.

In North and South America, natural gas production and consumption are largely balanced.

In terms of **coal production and consumption**, Asia and the Pacific make up an overwhelming part of the world.

Africa has so far accounted for a very moderate share of coal consumption. But what happens when China builds so many coal-fired power plants in Africa? The Chinese, of course, will also bring the coal.

1.2   Decarbonization in EU documents

In the following, the most important paragraphs and formulations of relevant EU documents are mentioned or quoted in a modular manner:

"Tightening up the existing EU Emissions Trading Scheme

The EU emissions trading system puts a price on greenhouse gases. So far, this has affected energy companies, energy-intensive industry and parts of aviation. In the past, too, the caps on the total emissions of individual sectors of the economy have been lowered every year.

In the future, emission rights are to be reduced even more sharply – by 2030 by a gradual 62 percent compared to 2005 (previously 43 percent). Efficient companies are to receive free **emission certificates** in the future. On the other hand, there will be cuts in the case of inefficient plants if those responsible do not implement efficiency measures.

So far, aviation and industrial sectors that are particularly in international competition have received free emission certificates. These are to be phased out gradually. Maritime shipping is also to be included in emissions trading from 2024.

Part of the revenue from the <u>EU emissions trading</u> scheme for energy, industry, aviation and shipping goes to the Innovation Fund, which aims to promote investment in climate-friendly technologies

New emissions trading scheme for transport and buildings

More greenhouse gases must also be reduced in transport and buildings. For this reason, a new emissions trading system for buildings, road transport and the use of fossil fuels in certain industrial sectors is to be created from 2027 – similar to the German fuel emissions trading system. Maritime shipping will also be included in emissions trading.

The CO_2 certificates are to be traded freely on the market – as in the previous European emissions trading system. There is no provision for free emission allowances.

Social Climate Fund for more climate protection measures and social balance

The revenues of the new emissions trading scheme for buildings and road transport of 65 billion euros are to be used to finance a new social climate fund from 2026 to 2032. In addition, the Member States will contribute their own budgets to the measures, so that a total of around 80 billion euros will be available for social compensation.

The fund is primarily intended to support measures in **more efficient buildings and lower-emission mobility,** and will mainly benefit lower-income households and small businesses. On a temporary basis, the Fund can also finance direct income support for particularly vulnerable households.

CO_2 border adjustment for competitive companies

As early as 2023, a CO_2 border adjustment mechanism will be introduced with a test phase of three years. A CO 2 price is to be levied for the electricity sector and selected goods imported into the EU, for example from the cement or steel, aluminum or fertilizer industries. The mechanism is intended to compensate European companies against companies from other economic areas that are not covered by EU emissions trading.

The aim is to ensure that Europe's ambitious climate policy does not lead to a shift of greenhouse gas emissions to other countries by relocating companies or production capacities. European companies should remain competitive. The CO_2 border adjustment mechanism is closely linked to the Emissions Trading Directive. It is intended to replace the allocation of free emission allowances by 2035.

Europe aims to become climate-neutral by 2050 and to reduce greenhouse gas emissions by at least 55 percent by 2030 compared to 1990 levels. The EU Climate Law sets out these targets in law for the first time. To this end, the EU Commission presented proposals for more than twelve legislative amendments in the summer of 2020. They are to be used to implement the new climate targets.

Other EU plans concern the expansion of renewable energies. This is because the expansion of solar and wind energy is to be massively accelerated throughout the EU. <u>By 2030, 45 percent of gross final consumption is to come from renewable energy</u>. This significantly increases the previous target of 32 percent. The preliminary agreement of the EU energy ministers provides for amendments to the EU directive to accelerate the expansion of renewable energies: In priority areas, approval procedures will be accelerated from 2023. There will only be a strategic environmental assessment at project and planning level. Renewable energies and the necessary grid infrastructure are recognized as an overriding public interest.

Approval procedures for <u>solar installations on buildings</u> and for <u>heat pumps</u> are to be shortened. The replacement of existing systems with newer and more powerful technology is to be simplified. German wind priority areas will be recognized as "go-to areas" at EU level and projects in such areas will be approved more quickly.

Increasing energy efficiency

In order to reduce energy consumption, reduce greenhouse gas emissions and fight energy poverty, EU countries need to use energy more efficiently. In the Energy Efficiency Directive, the Commission has proposed a higher annual target for energy savings at EU level. This would significantly increase the existing EU-wide savings target. Compared to the expected development of consumption by 2030, <u>primary and final energy consumption in the EU must fall by nine percent</u> *(these terms are explained below).*

From 2035 only, CO 2-free new cars

From 2035, newly registered vehicles in the EU will no longer be allowed to emit CO_2. This is because the fleet limits for passenger cars are to be reduced to zero by 2035. The EU member states have already reached a final agreement on this."

End of a long quote.

1.3 Emission certificates as letters of indulgence?

In terms of their felt-like goals and their possible, rather unmistakable effects, the emission certificates are strongly reminiscent of letters of indulgence, which were introduced in 1470 and then banned in 1562, under penalty of excommunication. Martin Luther himself has a clear position in his 95 theses against it. Have the EU representatives forgotten history, which should be rather instructive?

A quote from [8]:

"As soon as the money rings in the box, the soul jumps to heaven!", a sentence that the Dominican monk, Johann Tetzel, an obvious opponent of Martin Luther, is said to have uttered. With these words, he gets to the heart of the absurdity of the sale of indulgences. In the

Middle Ages, the church promised grace in exchange for money, salvation of souls with letter and seal. After the French cathedral dean Raimund Peraudi organized a papal indulgence in Rome around 1470 to restore the dilapidated cathedral of Saintes, the widespread capitalization and professionalization of the sale of indulgences began in the 15th century.

The sale of indulgences with his famous letter has been banned in the Roman Catholic Church since 1562 and has been punishable by excommunication since 1567, but the indulgence still exists today – only without cash flow. Today, thanks to church tax and donations, the big money comes directly to the Account.

And what is to be done with all the carbon dioxide emission certificates?

<u>Primary and final energy consumption</u> in the EU must be reduced by nine percent, but then comes the hammer, with ideal, rigid solutions, far away from technological openness and beyond the laws of physics and the experience of engineering:

Only <u>electric cars, Europe "climate neutral" by 2050,</u> i.e. with at least <u>ZERO carbon dioxide emissions.</u>

1.4   The path from primary energy to end and user

The world's population increased from 1 billion inhabitants (1800), which is no more than Europe means today (2018 – 746 million [9]), to almost 8 billion (2023). If the steep increase in <u>primary energy consumption</u> is analyzed, the first explanation is the increase in the number of the *world's population*. However, this is followed by those needs that have not yet been known to so many people: *food, electricity, house, light, heat and air conditioning, mobility with cars, planes and ships.*

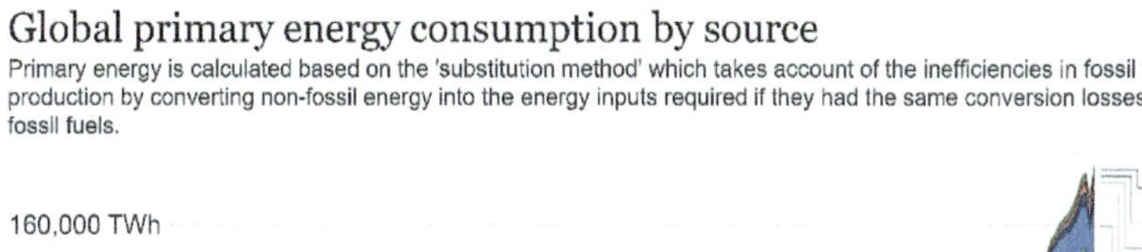

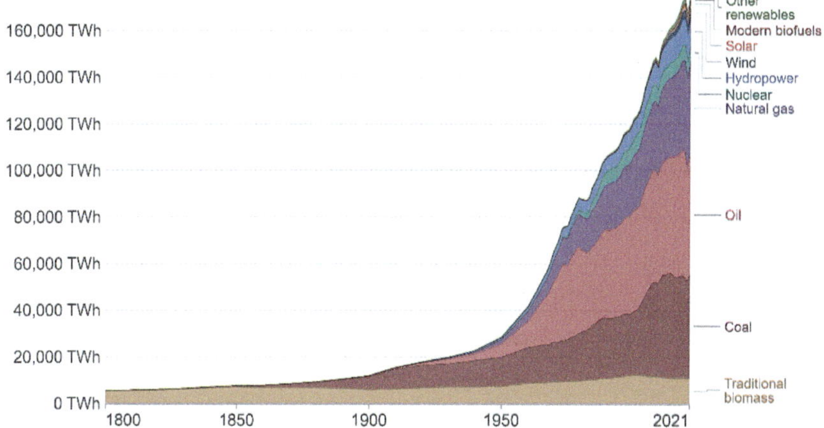

Figure 1.2 Increase in primary energy consumption in the world from 1800 to 2022 with the respective fossil (coal, gas, oil) and renewable (wind, solar, biomass) part [10]

The rapid increase in primary energy consumption in Figure 1.2 clearly identifies its main contributors: *coal, crude oil and natural gas*, and partly hydropower (power). The use of <u>wind</u> and <u>sun</u> in **wind turbines and** via **photovoltaic panels** shows two clear trends:

- <u>their contribution </u>is still, for decades, negligible on a global scale: in China, for example, in 2021, around 23,936 terawatt-hour TWh (one TWh corresponds to one million kilowatt hours – kWh) of primary energy demand was covered by coal, 1715 TWh by wind turbines and only 856 by solar plants, which corresponds to a ratio between wind and photovoltaics to energy from coal of **5.9%**. And China in particular is the largest user and importer of electric cars. How do they produce the electricity? With coal?

In the U.S., coal contributes to "only" 8,937 TWh, wind to 1,004 TWh, and solar to 433 TWh. The share of wind and solar compared to coal is about the same as in China: **5.6%**.

- An <u>increase in the contribution</u> of wind and solar energy in recent decades can hardly be seen from a global perspective in Figure 1.2 registered coal, on the other hand, has seen a significant increase, especially in the last 10 years.

Conclusion: *a significant increase in the contribution of renewable energies to the world's primary energy demand is absolutely necessary to protect the climate from carbon dioxide-induced warming! However, this cannot be achieved by forcing the introduction of electric cars that consume coal-fired electricity!*

<u>Electrical energy</u>, in turn, is used as a <u>background</u> in statistics of all kinds in order to highlight the contribution of renewable energy components from wind and sun. However, this background always remains quite obscure: if only electrical <u>energy</u> is meant, then the percentage value of the share of photovoltaics becomes important. <u>If the total primary energy demand</u> is meant, then the same proportion of photovoltaics makes a percentage fall!

In this context, it is worth noting that **13% of the world's population has no access to electricity at all.** Access to electricity, in this case, is defined as the ability *to use electric light, charge electrical appliances such as smartphones, and run a radio for 4 hours a day.*

Electrical energy, on the other hand, represents **20.4% of the world's energy consumption** [11].

Is <u>it</u> possible to compare primary energy consumption with the <u>consumption of electrical energy</u>?

From this point on, clear categories must be defined

Definition

Primary energy: a cumbersome formulation, as in some books, just for the sake of definition, is dispensed with here, it can very quickly become misleading and is often contradictory, depending on the author. So, at this point, just

examples of primary energy: wood, coal, oil, natural gas, solar (no photovoltaic system), wind (no wind turbine), water.

The primary energy is used for use in

> **Definition**
>
> **Secondary energy,** causing efficiencies, is stored in tanks or batteries.

As a result of the transport of the stored form of energy, *whether by ship, rail or cable,* to the electricity at the socket at home or in the tank of the vehicle, *losses* arise that reduce the stored value – this is why a final energy is defined.

> **Definition**
>
> **Final energy** is the energy used by industry, households, commerce and services. Primary energy sources can be used for this purpose, or secondary forms of energy, as an example of electric current.

> **Definition**
>
> **Useful energy** is the effective power of an electrical appliance for a certain period of time, or for thermal systems, such as heaters or combustion engines, the heat actually transferred, or the work actually performed. Thermal, mechanical, loss- and friction-related efficiencies also appear in this process.

Example: hydrogen

Primary energy: *water + sun / wind*

Transformation: Splitting of water by electrolysis or in the fuel cell with energy from solar radiation, converted into electricity by means of photovoltaics.

Another method to obtain hydrogen: by wind energy, converted into electricity by means of wind turbines.

Because the sun does not shine all the time and the wind does not blow all the time, the full load number was introduced, giving as a measure of the degree of utilization, in comparison with the planned, maximum power, usually over a year. Offshore wind power can achieve a full load number of up to 51.4%, while onshore wind turbines can reach up to 41.2%.

It should be noted that there is a fundamental difference between the full-load hours (duration of the "virtual" function at full load compared to the 8760 hours of a year) and the efficiency (effectively used by the energy used, in the case of energy conversion, for example from the heat supplied from a fuel to the output of work in an internal combustion engine). A few full-load hours can be <u>extensively</u>, i.e. by the number of wind turbines or by die area of the solar panels. The efficiency during the conversion of a fuel from heat to work should be <u>intensive</u>, i.e. through changes in the process, as far as this is possible.

"Electrolysis from water requires 3-4 times more energy than direct electrolysis

power generation", for electric motors, which some "experts" consult, is therefore only a comparison of apples with oranges? The sun's radiation, for example, is inexhaustible. More energy means more radiant area. In the end, the efficiency is shown by the price.

Secondary energy: *Hydrogen in the tanks, for example in Patagonia. Transport to Europe usually results in a reduction in energy, which is due to the energy consumption of the means of transport itself (ship, plane, truck).*

Final energy: *the energy of the hydrogen in the tank of the automobile.*

Useful energy: *the effective operation of the internal combustion engine that drives the car. Between the energy of the hydrogen (usually referred to as the calorific value [megajoules/kg of hydrogen]) and the effective work [kJ or, converted, kWh], a thermal efficiency (depending on the process control) usually has to be considered.*

References for Chapter 1

[1] https://www.umweltbundesamt.de/themen/klima-energie/klimawandel/klima-treibhauseffekt

[2] https://www.leopoldina.org/uploads/tx_leopublication/2021_Factsheet_Klimawandel_web_01.pdf

[3] V. Masson-Delmotte, P. Zhai, A. Pirani, S. L. Connors, C. Péan, S. Berger, N. Caud, Y. Chen, L. Goldfarb, M. I. Gomis, M. Huang, K. Leitzell, E. Lonnoy, J.B.R. Matthews, T. K. Maycock, T. Waterfield, O. Yelekçi, R. Yu, B. Zhou (Hrsg.):
Climate Change 2021: The Physical Science Basis. Contribution of Working Group I to the Sixth Assessment Report of the Intergovernmental Panel on Climate Change. 6. Auflage. Intergovernmental Panel on Climate Change, Genf 2021, ISBN 978-92-9169-

[4] Jim Skea et al. 2022: *Climate Change 2022: Mitigation of Climate Change. Summary for Policymakers* (Memento vom 7. August 2022 im *Internet Archive*). Sechster Sachstandsbericht des IPCC

[5] https://de.statista.com

[6] /doi/10.1126/science.aau5153

[7] DOI: 10.1399/ NuL.2021.12.01

[8] https://qiio.de/die-kapitalisierung-des-seelenheils

[9] Deutsche Stiftung Weltbevölkerung dsw.org/.../Datenreport-2019

[10] https://ourworldindata.org

[11] https://yearbook.enerdata.net/electricity

Chapter 2

Battery-powered electric cars versus combustion engines with climate-friendly fuels: complete better than delete

One hundred percent electric cars with batteries in Europe in exactly 12 years, zero "burners", according to European laws, regardless of whether they can also burn carbon-neutral fuels. And what is true for Europe is usually generalized for the whole globe. However, there are 1.4 billion cars in the world, of which 10 million, i.e., less than one percent, are electric cars.

From naught to sixty, where will the electricity come from?

In the former Canton Zone (Guanzhou-Shenzhen) with 65 million people, electric cars with their zero emissions are, theoretically, a necessity: if it weren't for the millions of two-stroke engines equipping two- and three-wheeled vehicles that spit out not only carbon dioxide but also unburned oil and hydrocarbons. The electricity for electric cars that drive through the metropolises is usually produced outside them. In China, 68% of the total primary energy comes from the 1077 coal-fired power plants, with 30 new ones added every year. India, also known for megacities where electric cars are also needed, currently produces more than 60% of its electricity from coal and natural gas.

Nevertheless, according to numerous information about practical tests, a mid-range car spends about 20-30 kWh / 100 km: that is exactly 4500 kWh with an average distance of 15000 km per year, leading to 30 kWh for every electric car, which corresponds to the peak performance of an offshore wind turbine that is stated everywhere. Do we build a wind turbine for every electric car?

The world-saving, very idealized scenario of electric cars with batteries, which refuel with electricity from photovoltaic or wind turbines, absolutely without combustion, absolutely emission-free, is neverthe-less vehemently defended by its proponents. But what do we do with the electricity produced in this way, if not necessary at the moment, store or forward it? To store it, you have to take the battery which weighs several tons, from Tokyo or Alaska to Patagonia or Kenya and then back again. To forward it you it need cables. Costs, infrastructure changes, energy losses in the cables enormous.

Fortunately, storing or forwarding have an alternative, which is on site: producing hydrogen, because packaging it in ammonia and bring-ing it to Europe, or hydrogenating fry fat and plant residues (SAF (Sustainable Aviation Fuels), HEFA (hydro processed esters and fatty acids), biokerosene remains as cereal rusts and algae, it remains the synthesis of hydrogen and carbon dioxide from the environment.

Last but not least, there remain two groundbreaking potentials: Meth-anol and Ethanol , from all spoiled fruits and plant remains, or from sugar cane.

The "combustion engine" is coming back, but with climate-friendly fuels, not to replace electric cars with batteries, but to supplement them. By recycling carbon dioxide in nature, such "burners", powered by synthetic and renewable fuels, do not cause climate change and are used in vehicles, on ships, in trucks in construction machinery.

2 Battery-powered electric cars versus combustion engines with climate-friendly fuels: complete better than delete

2.1 Electric cars, emission-free: local, but not global

The European Union's law on the *exclusive registration of electric cars*, which will apply to EU countries from 2035, could, possibly, eventually apply worldwide, according to the hope of the proponents.

But let's see what the reality looks like:

The number of vehicles on earth already exceeds **1.4 billion units**, out of about **8 billion people**, although their distribution seems to be completely inhomogeneous: very many people and vehicles in Western Europe (602 out of 1000 people own a car), in Far East Asia and in the USA (830/1000), but hardly in Africa (*44/1000*), where no less than 17% of the world's population lives! [1]

There are currently just over **10 million electric cars** worldwide, which corresponds to an almost insignificant share of 0.71% of the total global vehicle fleet.

© The Author(s), under exclusive license to
Springer-Verlag GmbH, DE, part of Springer Nature 2024
C. Stan, *Energy Scenarios for the Future*,
https://doi.org/10.1007/978-3-662-69687-3_2

The annual production of automobiles in the world, whether with gasoline or diesel engines, with fuel cell electric motors, with batteries or plug-in type, was **55.8 million** units in 2020 alone. Toyota produces over 10 million cars a year, VW almost 9 million. Tesla has yet to reach more than 0.9 million electric cars in 2021, despite spectacular increases!

However, a fundamental problem arises: Where will the electrical energy come from if all cars have to become electric? The share of electricity of 20.4 percent of the total energy used is currently already used for electrical and electronic consumers, such as computers, lighting, *vacuum cleaners, air conditioners, trams and street lighting*. With the exception of Brazil and Sweden, which are based on local specifics, the renewable forms of energy used to generate this electricity (*hydropower, solar and wind, biomass, garbage, geothermal energy*) remained at values around 20% as shown in Figure 2.1.

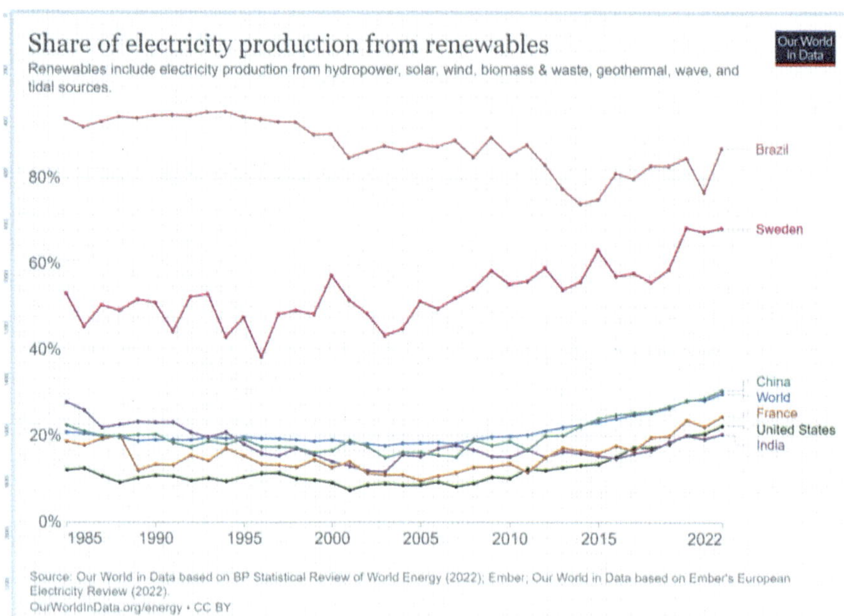

Figure 2.1 Share of electricity renewable forms of energy (hydropower, solar and wind, biomass, waste, geothermal energy (*Source: Our World in Data/energy*)

What would then be the impact of the explosive growth of the electric propulsion of automobiles on the electricity consumed worldwide? By 2030, the global electric car fleet is expected to grow to **116 million** *(Statista, June 2020)*.

In the same context, however, the world production of cars with all configurations of propulsion systems grew and continues to grow enormously: in 2001, China ranked 14th among car manufacturers with 0.7 million vehicles per year. Japan, which was in first place that year, had produced 8 million cars, followed by Germany with 5.3 million. Ten years later, by the time of 15 years, China had 8 million cars conquered the first place in the world and now reaches no less than **25 million**!

However, the other side of the coin is that only 4% of Chinese cars are exported to Europe, America and other continents! *Will it be similar with the electric cars, which have currently conquered a third of the Chinese market (2023)?*

What is absolutely remarkable, however, is that the hierarchy of producing countries is also changing a lot: India was in 15th place in the manufacturer ranking in 2001 with 0.65 million cars and ten years later reached sixth place with 3.6 million. And more recently, India, with 6.2 million cars, ranks ahead of countries with a strong automotive industry, such as Germany (currently 6.2 million), Spain (2.6 million), France (2.4 million) or the United Kingdom (1.7 million).

China and India are also the most populous countries in the world, with practically half of the world's inhabitants. The highest birth rate in the world is recorded in India. A correlation between the number of inhabitants and their transport needs would be worth mentioning at least.

Many people, for example in India, Congo or Brazil, still simply lack the basic means of subsistence: *heat, electricity and food.* And where should they seek shelter faster than within or around large swarms of people, in metropolises or in megacities? **32 million people** are counted in **Delhi**, Mumbai 21 million, Calcutta 15 million. On the other hand, in China, **Guangzhou-Shenzhen has 65** million inhabitants (like the *entire population of France),* Shanghai 39 million, Beijing "barely" 20 million, that is, as much as the largest city in the USA, New York City!

The traffic of cars with internal combustion engines, through cities of this size, can endanger people's lives, the health of animals, but also the growth of grass and trees.

District power plants, on the other hand, are usually located outside densely populated centers. But this is not possible with vehicles, which have to cross cities far and wide, on millions of thick and thin arteries.

The solution, then, remains to allow neither high concentrations of carbon dioxide, nor nitrogen oxides, nor fine particles that can be carcinogenic, in the midst of these poisoned Babylonian cities.

The emission-free electric car is an unavoidable solution under such conditions.

In China alone, no less than 2.7 million cars with electric motor and batteries were sold in 2021 *alone (Statista, January 2022),* while their sales stagnated at around 0.3 million per year in the United States and 0.88 million vehicles in the European Union as a whole!

And what about the U.S.? A major German car manufacturer is currently offering the new model of a luxury sedan with an electric motor of 400 kW (544 hp) for the EU. The maximum torque is 745 Nm. The battery has a capacity of 105 kWh (kilowatt-hours) and provides a range of around 600 kilometers. However, such a car has a total weight of at least 3.2 tons.

It's just that the Americans in the USA, i.e., those who advertise their **Tesla** cars a lot in the world, from Europe to Asia to Australia, don't request electric cars from the German manufacturer, but something super modern with a piston engine. Thus, the famous German manufacturer offers the Americans a variant of the same sedan, but with an internal combustion engine – *not with 4 cylinders, not even with 6,* *but with 8, which brings the displacement to 4.4 liters*! The engine develops a power equivalent to that of its electric pedant, but which must remain in Europe.

We want zero-emission electric cars in the EU and even in the rest of the world. An honorable and tempting commitment! But the reality is harsh:

84.3% of global energy currently comes from fossil fuels – oil (33.1%), coal (27%), natural gas (24.3%). Nuclear energy generates 4.3%, hydraulic energy 6.4% as shown in Figure 2.2.

2.2 Only electric cars for the EU, and for the rest of the world?

Wind energy, onshore and offshore, i.e., on earth and oceans of the world, implemented in countless wind turbines, provides no more than 2.2%. And photovoltaic systems, which are captured in fields, on houses, on balconies and even at sea and transformed into huge panels, has a currently negligible percentage of only 1.1%. The sun doesn't shine always, still everywhere.

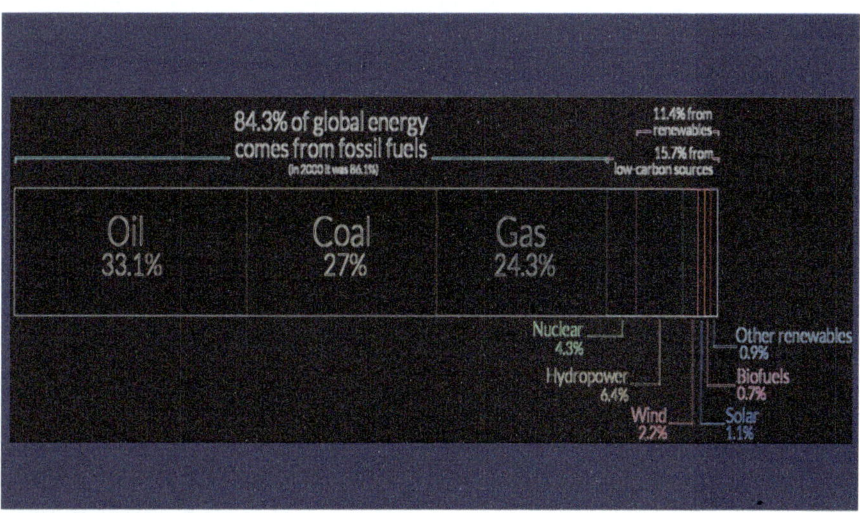

Figure 2.2 Shares Primary energy currently comes mainly from fossil fuels *(Source: Our World in Data*

It is true that the electricity for electric cars, which mainly drive through large metropolises, is usually produced outside them.

The problem, however, is that _68 percent of the primary energy in China,_ including the electricity for electric cars, comes from coal-fired power plants!

China, which sells by far the most electric cars in the world, currently _(July 2020)_ operates no less than 1077 coal-fired power plants, between 20 and 30 new power plants are commissioned annually! And this is a sensitive point of energy management in China, between electromobility and the energy required for it: Although 68% of electricity generation is based on coal, on the other hand, only 55% of the energy that China needs in all industries is produced on the basis of coal! Gas and oil are mainly used for heating and industry.

The higher the number of electric cars in China, the more the demand for electricity will increase, which is usually accompanied by an increase in the amount of coal used to produce them.

And China, the world's largest user, manufacturer and importer of all-electric, battery-powered cars, also holds the record for carbon dioxide emissions: one-third of global emissions! The main cause of these emissions is the burning of coal.

Share of electricity production from fossil fuels

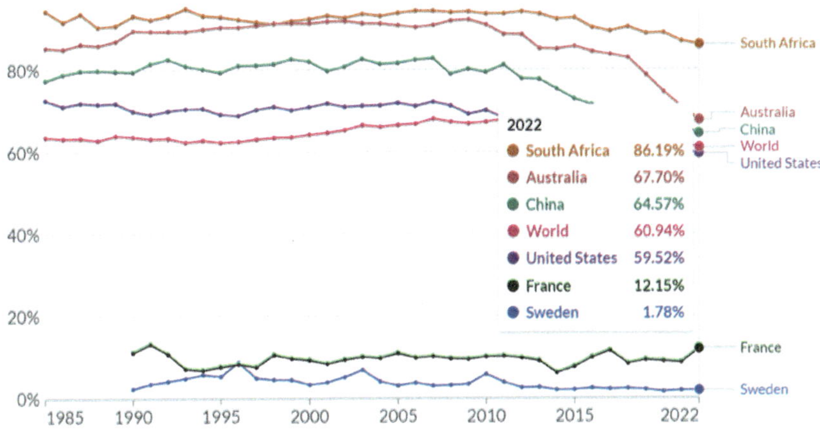

Figure 2.3 Worldwide electricity generated from fossil fuels (*coal, gas, oil*) from 1990 to 2022: (*Source: Our World in Data*)

Coal production

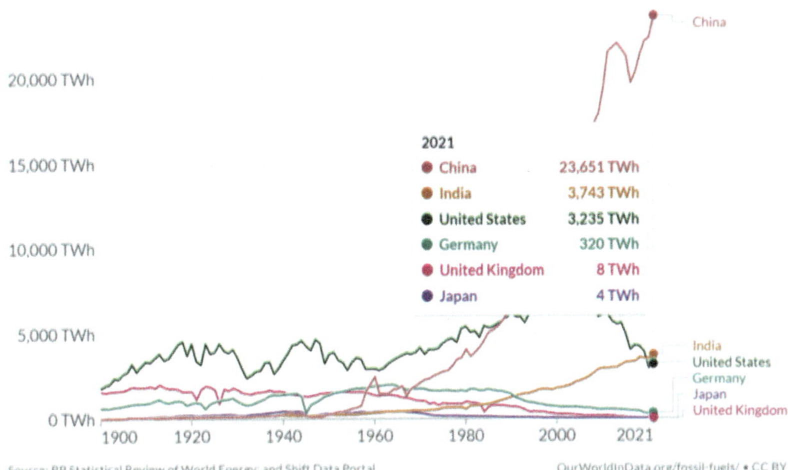

Figure 2.4 World coal production from 1990 to 2022: China, 23,651 TWh, more than <u>India, USA and Germany</u> together (*Source: Our World in Data*)

India ranks third in terms of carbon dioxide emissions (7.3%), behind the US (13.5%). Due to the different social structure in India than in China, electric vehicles have not yet become large consumers of electricity there. However, **India currently produces more than 60% of its electricity from coal and natural gas.**

What good is it to us to note that Belgium, Austria, Sweden, Portugal, soon Italy, the Netherlands and other European countries have abandoned coal-fired power plants or are abandoning them altogether, while electric cars are on the road in China, India and the rest of the world?

Admittedly, electric cars have different requirements in Europe, to some extent:

Norwegians are at the forefront of electric mobility, with 87% of registrations of all-electric and plug-in vehicles in 2021. There are suitable reasons for this electrification of mobility: there is no nuclear power plant or coal-fired power plant in Norway, instead it operates no less than 1690 hydroelectric power plants, which cover **over 90% of the country's total energy needs, including electricity**.

The Norwegian state certainly has a lot of oil and natural gas, but prefers to export it elsewhere: Norway ranks 2nd in Europe and 12th in the world in oil exports. In addition, the 9th place in the world and 2nd place in Europe in the export of natural gas. Let the importing countries burn this oil and natural gas in their own air! As if carbon dioxide emissions from China or Poland, through the skies, could be stopped at Norway's borders!

The French operate 56 nuclear power plants, which provide a share of electricity equivalent to that of coal in China, i.e. 67-68%. Its hydropower plants contribute 13% to this, while wind energy accounts for 8%. So, it's no wonder that the French are once again at the forefront of electromobility.

The Poles, on the other hand, get 76% of their electricity from coal, in the middle of the EU!

The Dutch, big proponents of electric mobility, produce electricity mainly from natural gas (42%) and coal (21%).

The reaction of car manufacturers, especially those that have been operating successfully on the national market and also worldwide for a long time, is strongly reminiscent of the strategy of very pragmatic folk heroes during the French Revolution:

We will have to sacrifice the king of automobile propulsion, that is, the internal combustion engine, we will also go through the electric revolution, trying to have minimal losses, but we will properly prepare the coronation of the emperor with many pistons, turbochargers and direct injection of renewable fuels!

On July 14, 1789, the French Revolution de facto began in Paris with the attack on the Bastille state prison. During the revolution, which lasted about ten years, the people turned the social order upside down: the prerogatives of the king and his followers had to give way to new rights for ordinary citizens. Moreover, in a trial before the National Convention, King Louis XVI was sentenced to death and executed on 21 January 1793. And then? Only 15 years after 1789, on December 2, 1804 in the Cathedral of Notre Dame de Paris, during a ceremony in the presence of the Pope of Rome, the former simple revolutionary player Napoleon Bonaparte, marching among the rank and file, proclaimed himself emperor! The story seems paradoxical, but it is very instructive:

At that time King Louis XVI was sentenced to death and executed by the impetuous France, by a revolutionary convention, similar to the parliament of the people's representatives in Brussels. The "combustor" is now, at the mercy of the EU representatives, to have the same fate. Napoleon was a supporter of the revolution at the time, but later he put on the imperial crown, for France and then for Europe! Greetings from Ferdinand Porsche, which is not the father of the electric car, but the representative of the very strong and famous "burner"!

Is the story of **electric cars** taking the same course? In addition to Tesla in the USA, the current revolution in electromobility has car manufacturers such as VW in Germany, but also the Chinese from

BYD, SAIC and GeelyVolvo or Nio, the French with Stellantis and Renault, the Japanese with Toyota and Nissan and countless start-up companies, for example Lucid Motors in the USA, in its ranks.

Next to them, however, slowly but surely, the pearls and precious stones of the future imperial crown appeared, the " **burners"** of the new kind:

Very trendy sedans with "burners" developed and launched in 2022, that is, only 13 years before the full electrification of the European market, such as the *Mercedes C300 4matic, the Genesis G70 2.0T Plus AWD, der Audi A4 45 TFSI quattro, the BMW 330i xDrive* have 2-litre four-cylinder petrol engines. The new Peugeot 308 offers 3 powertrain variants, two with combustion engine (diesel and petrol) and a plug-in.

In the SUV category, more gems appeared with "combustion engines", such as Audi Q5 45TFSI quattro, BMW X3 xDrive301, Mercedes GLC 300 4Matic, all with gasoline, 2-liter four-cylinder engines, similar to those previously mentioned for sedans. Range Rover offers no less than 5 piston engine variants in the new 5 Series – three diesel and two petrol.

The combinations between combustion engines and electric motors in hybrid propulsion systems are diverse and countless, in most cases the combustion engines also generate electricity and also participate in the propulsion.

An engine that passes through all torque and speed ranges is more difficult to optimize in terms of consumption and emissions than an engine that operates at a fixed point, only as a power generator. The concept is not exactly new: it was successfully tested in the 2000s in the Citroen Saxo and later in the BMW i8. However, it did not enjoy too much acceptance at the time. At the moment, however, Nissan is bringing it back to the market with the new X-Trail model: The power generation on board is taken over by a 1.5-liter gasoline engine that develops an output of 116 kilowatts (158 hp), the drive is provided by a 150-kilowatt electric motor.

Of course, the electric drive of cars will gain a lot of ground internationally, even though coal is burned in power plants for its benefit. But the fronts are slowly clarifying:

Electric cars will be indispensable in urban areas, but not for the rest of the world.

Tesla's new electric competitor is Lucid Motors. The recently launched Lucid Air model has a power of 794 kilowatts (1080 hp), a battery of 113 kilowatt hours, but also a total weight of 2.4 tons. The company's next model will be Lucid Gravity Figure 2.5, an electric SUV with similar technical characteristics.

Figure 2.5 Electric Cars with Batteries: Lucid Gravity und Fiat 500 (*Source: Lucid und Fiat)*

For comparison, an urban electric car, the Fiat 500, has a 70 kW (95 hp) electric motor, which is less than a tenth of Lucid's power, but sufficient for which a 23.8 kilowatt-hour battery is, with a satisfactory range in normal city traffic. However, a Fiat 500 weighs a ton less than its big American brother, the Lucid, which seems very favorable for the lively and very turbulent movement of sheet metal avalanches through Rome or Paris.

2.3   "Burner" with hydrogen for the big world, electric car for the city

Hydrogen is the first to be climate-friendly: its thermal conversion by combustion in a "burner", in a mixture with the air from the environment, basically leads to water, occasionally, even slightly, to nitrogen oxides.

In a fuel cell, which produces electricity on board vehicles for electric propulsion, the chemical reaction also results in water. That sounds forward-looking for a climate-friendly fuel. Unfortunately, hydrogen does not exist individually in nature, it must be extracted from other substances.

One possibility is the electrolysis of water as shown in Figure 2.6. Unfortunately, as already mentioned, only 2% of all hydrogen produced on a global scale comes from water electrolysis. Industrial, global hydrogen production has been state-of-the-art for one hundred years, especially in the fertilizer industry and fuel refineries. For these applications, however, hydrogen is derived from natural gas (38%) and from other hydrocarbons, such as heating oils and heavy fuel oils, 10% even from coal. The conversion of such hydrocarbons releases carbon dioxide in addition to hydrogen. This is independent of the energy that has to be used in the conversion process to hydrogen.

Fortunately, we owe electrolysis directly from water to the English scientist Michael Faraday (1791 – 1867), who essentially derived or built the laws and laboratory facilities for this purpose

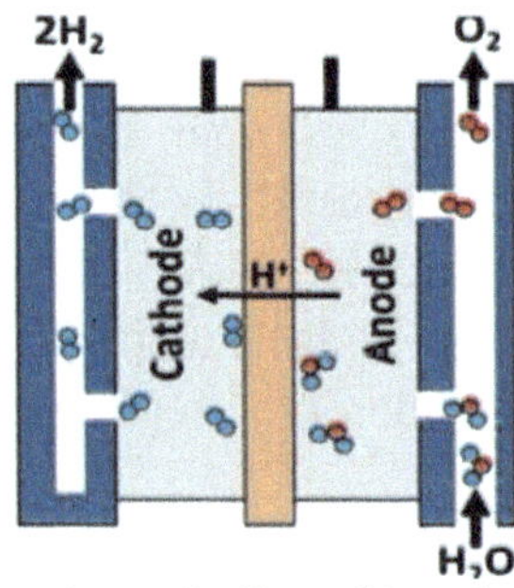

Figure 2.6 Water electrolysis schematically and in a fuel cell with "reversed" reaction sequence in a current plant (source: US Department of Energy [2]).

Water electrolysis for the production of hydrogen appears to be absolutely advantageous in view of climate protection, because the other reaction product, besides hydrogen, is simply <u>oxygen</u>.

A "problem" that is sometimes criticized by "energy experts" is that the manufacturing process itself requires more energy than that contained in the hydrogen produced. It is very easy to calculate an "energetic efficiency" (but this is false, because the energy for water electrolysis, usable for the production of hydrogen does not come from a tank, container or storage facility of any kind, but from the <u>sun's rays</u>, photovoltaic systems and <u>wind energy</u>, using wind turbines). We only have occasional rays of sunshine and wind (sun/shade or storm/win, but for numerous spaces and times, from Alaska via China to Tierra del Fuego and Australia and from prehistoric humans to Homo sapiens.

Such plants and modules, of all sizes, have recently sprung up from the ground, for both large and small quantities of hydrogen. Only the energy source (*sun rays or wind*) and water are needed.

And when it comes to "energy efficiencies" as a major disadvantage of hydrogen electrolysis, we can also turn the tables!

Although the process itself requires less energy for hydrogen in direct electrical energy production than in water electrolysis, whether this electrical energy is used for storage in batteries or for onward transport to the consumer, we simply ask ourselves the question: Where is the consumer? Should all electric cars be moved to Patagonia or Egypt to recharge their batteries?

There is still the option of transporting the large and tons of batteries full of electrical energy from Patagonia to New York by tugboat. The tugboats, in turn, consume heating oil or heavy fuel oil in their engines. And the road is long, very long, the one in the chain, global "energetic efficiency" becomes a disaster.

Electrical cables, via high-voltage lines from Ushuaia/Tierra del Fuego to New York/USA, i.e., over 10,628 km, is also not an alternative: there is a transmission loss *or* **grid loss,** as the difference between the electrical power generated in the power plant and the electrical power used at the grid connection point of the consumers. The current flowing over lines causes transmission losses mainly due to electrical resistance, which leads to heat dissipation, as energy loss. A transmission loss of approx. 3 percent per 1000 kilometers is expected. [3].

On the other hand, hydrogen, as an energy storage device, does not require heavy batteries or infinitely long transmission cables, but only ordinary tanks or, more recently, chemical storage.

The places of production of electrical energy, for example for hydrogen production, from wind, sun and water are not uniformly distributed on Earth, but in regions that have a lot of wind or sun most of the time:

Jiuquan Wind Power Base, in China, as an example, is the world's largest wind turbine, with a planned capacity of **20 gigawatts** (GW) (currently: 6 GW)

Jaisalmer, in India, is a wind turbine with a planned capacity of **1.6 gigawatts** (2012), like that of *Alta Wind Energy Center* in California.

1 gigawatt each (2018) was generated by offshore wind turbines in *Norway* and the *Walney wind farm* in Great Britain (2018).

One of the world's largest photovoltaic systems

Pavagada Solar Park, India, is planned for a maximum capacity of **2.05 gigawatts**, gained on an area of solar panels of **53 km²**

For comparison, the *Huangh He Hydropower Park hydropower plant in* Hainan, China, generated **2.2 gigawatts** (2021).

The average efficiency of a photovoltaic system (expressed as a percentage in full-load hours) is only 13.7%, as long as it operates in summer and winter, in sun and shade. Wind turbines, on the other hand, reach between 16% and 57%. Such plants, which use wind and solar rays, are essential for the production of hydrogen,

even if they only work intermittently or only at low power: hydrogen can always be stored in any quantity!

For transport to Europe, the USA or China, the hydrogen must then be "packaged". Hydrogen does not necessarily have to flow through pipelines, like natural gas through "Nord stream" pipelines, but can be stored cryogenically, similar to LPG (Liquefied Petroleum Gas)

Hydrogen has the smallest molecular mass of all elements, and thus the lowest density, which is about 15 times lower than that of atmospheric air, with comparable values for pressure and temperature [4]. According to the equation of state of ideal gases, to increase the density, either the pressure can be increased or the temperature lowered. *Toyota, Hyundai* and other vehicle manufacturers have implemented pressure accumulators for use in automobiles with fuel cells at pressures of 300, 500, 600 or even 900 bar. As a comparison, in the tires of a mid-cassette car, the pressure is only 2.5 bar.

BMW has opted for cryogenic storage, at minus 253 degrees Celsius, in which the phase of hydrogen changes from gaseous to liquid, but at densities around a tenth of that of gasoline. However, the calorific value of hydrogen is 3 times higher than that of petrol. Hydrogen storage under pressure "costs" 12% of the energy of the same hydrogen mass, while cryogenic storage costs as much as 30%.

An alternative solution is to <u>store hydrogen in accumulators</u> with a metallic structure, which can absorb the hydrogen into pores, with the chemical compounds being rather weak. These can be dissolved by increasing the temperature.

One step further is the solidification of the hydrogen compounds, by chemical compounds, for example as <u>ammonia,</u> by the classic *Haber-Bosch process*, at 300 bar and 450 degrees Celsius. The ammonia can be liquefied at minus 33 degrees Celsius. A similar result is achieved at pressures above 9 bar. In order to release the hydrogen after transport, for example by ship or truck, it is sufficient to heat it at around 700 degrees Celsius. Liquid ammonia has a density comparable to that of gasoline, but its calorific value is only half that. The ammonia could also be burned in a conventional piston engine, on the consumer side. However, the problem could possibly be the formation of nitrous oxide.

2.4   „Burner" with alternative, climate-friendly fuels

LOHC (Liquid Organic Hydrogen Carriers)

These are chemical structures with carbon chains that are connected in nets, similar to wire meshes, with hydrogen atoms floating through meshes. The hydrogen to be transported can be "bound" with the help of a catalyst using a newer technology at 30-50 bar and 150-200 degrees Celsius and then, after transport, "set" by heating it to over 250 degrees Celsius.

The hydrogen can be used either in fuel cells or in piston engines. The reaction product is the same in both cases: water.

The advantages and disadvantages of using hydrogen in internal combustion engines compared to the "cold" chemical reaction in fuel cells can be deduced in base on following criteria the technical effort, the mass and the volume of such a system on board the vehicle, the efficiency in relation to each functional module and the system itself, and, finally, the price [5], [6].

Methanol und Ethanol

They are obtained from two groups of raw materials: *starch and sugar*, which come from plants or plant residues. In Brazil, sugar cane is used to produce ethanol, in the United States of America corn, in Europe sugar beet and partly wheat, in Asia cassava. New sources of raw materials for ethanol production worldwide include algae and cellulose from the paper- and wood-processing industries, as well as plant waste. In 2003, after a long period of positive experience with ethanol-adapted cars, followed by a sharp slowdown with economic causes, a car with a "flex-fuel" engine (*VW Gol 1.6 Total Flex*) adapted to variable mixtures of gasoline and ethanol from 0% to 100% was developed and approved in Brazil. Seven years later, the brands ***Chevrolet, Fiat, Ford, Peugeot, Renault, Volkswagen, Honda, Mitsubishi, Toyota, Citroen, Nissan and Kia*** entered the

Brazilian market with flex-fuel cars. A flex-fuel record of 94% of all new registrations was achieved!

Currently, there are almost 30 million flex-fuel vehicles in Brazil. The country itself has 355 million hectares of arable land, of which 72 million hectares have been cultivated so far. Sugar cane is grown on only 2% of arable land, and ethanol production requires about half that. Brazilian scientists claim that sugar cane production can be increased 30 times without harming the environment and without jeopardizing food production. Productivity is up to 8,000 liters of ethanol per hectare, at a price of 22 US cents per liter. The ratio between the energy consumption for growing crops/extracting ethanol and the energy actually used in the car is 1:10 - 99.7% of Brazil's sugar cane plantations are located on plains in the southeastern region of São Paulo, at least 2,000 kilometers from the Amazon rainforest, where the climate would not be favorable for sugar cane cultivation.

In the U.S., ethanol is produced primarily from grains and corn, on 10 million hectares, or 3.7% of arable land, and yields up to 4,000 liters of ethanol per hectare, half of what it is sugar cane production in Brazil. The energy balance is only 1:1.3 to 1:1.6 (Brazil 1:10) The basic price is 35 cents/liter.

Ford, Chrysler and GM build flex-fuel engines for their entire range of vehicles – from sedans to SUVs. There are currently more than 10 million flex-fuel vehicles on the roads.

Each spark-ignition engine can be switched from gasoline combustion to ethanol or methanol combustion, or variable mixtures of all 3 components. The results of many years of research both in our own institute, in the FTZ (*Research and Transfer Center Zwickau/Saxony, Germany*), as well as in many research centers of the automotive industry, with large and small engines, are very promising. A short summary is instructive:

- Engine torque increases by 10-15%, a result that Porsche and BMW have confirmed through their own research over the past 40 years.

- Pollutant emissions are significantly reduced in all applications.

- Methanol consumption increases by 2.2 times, ethanol consumption by 1.6 times. All these values are absolutely explainable: gasoline contains only carbon and hydrogen in its structure. However, ethanol also contains a proportion of oxygen. Alcohols, such as methanol and ethanol, with their own oxygen in the molecule, therefore require less oxygen from the ambient air. For a given amount of air in the cylinders of a given engine, therefore, in order to achieve the optimal combustion reaction, a chemically correct air/fuel ratio must be achieved: if the ratio is lower than that of gasoline, either less air or more fuel must be dosed. If the amount of air is given by the geometrical cylinder content, then only the second variant remains, i.e. more fuel.

An increase in fuel consumption does not mean a reduction in thermal efficiency, but is only determined by the proportions of air and fuel in the chemically exact reaction. On the other hand, combustion with its own oxygen in the structure of an alcohol is faster than with gasoline, which justifies an increase in engine torque even by the same amount of air in the cylinders as in gasoline injection. Therefore, switching from a gasoline engine to running on ethanol or methanol is usually simple: the injection system is adjusted to a higher dose of fuel, the ignition angle is adjusted.

On some engines, plastic pipes or gaskets still need to be replaced, but the problem is practically overcome, modern serial engines do not have any disadvantages of this type.

Synthetic fuels - synfuel, sun fuel, designer fuel (recently also e-fuels, because they are produced with electricity, from water and carbon dioxide)

are increasingly expressing a new trend in the research and development of fuels whose molecular structure can be specifically "constructed". Some features are worth mentioning:

Nature's renewable raw materials are inexhaustible: the *sun's rays, wind power, energy in plant residues, recyclable waste* from agriculture, the timber industry or the chemical industry.

The use of carbon dioxide from the ambient air or that which is produced in large quantities by blast furnaces, thermal and power plants, factories and plants is increasing.

Manufacturing with a minimum of energy, or with <u>electricity from wind and photovoltaic systems in favorable regions</u> where the consumption of solar or wind energy, which is infinitely available, does not play a primary role - more energy in this case only means more functional life or installation space.

Adapting such a fuel to the requirements of the combustion process in a specific engine or chemical reaction in a particular category of fuel cells is common, but also efficient.

The interaction with the environment is largely climate-friendly.

Of all the <u>non-synthetic fuels</u> listed so far, ethanol, methanol and dimethyl ether (which is extracted from wood residues) best meet most of these criteria.

Synthetic fuels from the synthesis of carbon dioxide and hydrogen,

which meet the above criteria are currently produced mainly by the <u>synthesis of carbon dioxide and hydrogen,</u> with the intermediate being <u>methanol</u>. The end products belong to the category of OME Ethers (polyoxymethylene dimethyl ether) with different molecular chain lengths: OME1, OME2, OME3, OME4, OME5, OME6.

As with ethanol, methanol or dimethyl ether, oxygen atoms in the molecule of an OME variant reduce the air required for the combustion reaction compared to hydrocarbons such as gasoline and diesel. For the same amount of air in the cylinders of an engine, the "consumption" of such a fuel increase with the proportion of oxygen it contains. However, this "consumption" has nothing to do with the thermal efficiency of the engine, but only with the proportions of the components in the chemical reaction during combustion. On the other hand, the torque of an engine running on such a fuel increases, similar to ethanol.

Porsche is currently building excellent electric cars that are at the forefront of international development in terms of power, torque, range, but also charging power.

On the other hand, 70% of those who have had "classic" combustion engines since Porsche was founded in 1931 are still on the road around the globe, most of them in perfect condition. Who would throw such a car in the garbage dump? Porsche wants to continue to secure the existence of these cars with "burners", but with fuels with no impact on the environment. However, such fuels are also envisaged for future models of the company, most of which will use high-efficiency and efficient internal combustion engines in the near and distant future.

Together with *Siemens Energy, ExxonMobil* and other high-profile partners, Porsche recently started producing "e Fuels" in *Patagonia, Chile*, in a project called "*Haru Oni*", which means "strong wind" in the native language of the local people. This strong wind blows more efficiently than anywhere else in the world into the rotor blades of a power-generating wind turbine, which is then used in a new "Silyzer" electrolysis system developed by Siemens, in which electrolytic reactions are based on an exchange of protons with polymer membranes, similar to the "reverse" reaction in fuel cells. Methanol is produced from the synthesis of hydrogen, with carbon dioxide, which in this case is not extracted from a blast furnace but from the ambient air. It can be used in engines as such, as mentioned above, or it can be converted into an OME fuel, which is also applicable in older automobiles with "combustion engines". The big advantage of using methanol or an e Fuel instead of hydrogen is the possibility of transport and storage in the vehicle at ambient pressure and temperature.

The production of methanol and OME fuels in Chile, in partnership with Porsche-Siemens & Co., started in 2022 with 130,000 liters per year and will reach 550 million liters in just four years!

The huge **container ship engines** are now using fuels of the same category. A prime example is Wartsila's four-stroke engines, which have been taken over from the direct-injection diesel engines previously fitted to these vessels. B&*W/MAN* has also developed engines with direct injection of methanol for use on ships, which are two-stroke engines because they have advantages over four-stroke engines in terms of weight and dimensions.

Both Wartsila's four-stroke engines and B&W/MAN's two-stroke engines use a combustion process that is fundamentally different from that of conventional compression-ignition (diesel). The direct injection of a small amount of diesel creates countless foci of ignition in the combustion chamber, through which the main amount of methanol is then injected, which, in the presence of air, rapidly ignites and burns simultaneously in all "activated" centers. This reduces both fuel consumption and nitrogen oxide emissions (*more detailed explanations in the book* [6]).

Four-stroke methanol direct-injection engines from Wärtsilä are used on the *ferry Stena Germanica* and the B&W/MAN two-stroke engines on the *Lindager* with 10,320 kW. The methanol for the "*Stena Germanica*" is produced from carbon dioxide that flows from blast furnaces used in steel production, which is emitted by Sweden's largest steel company SSAB in Luleå.

The Danish container shipping company *Möller-Maersk* has also ordered eight large container ships, which are to be operated with "green" methanol from 2024. Each of these vessels has a capacity of 16,000 standard containers (TEU). These ships are expected to reduce their carbon dioxide emissions by one million tons per year (from 33 million tons in 2020). In addition, the combustion of methanol with air produces significantly fewer particles and pollutants than the combustion of diesel or heating oil. The new vessels will be built by Hyundai Heavy Industries in collaboration with marine engine manufacturer MAN.

Similar motors are used to power generators in power and nuclear power plants, as well as in hospitals and banks.

Air travel could also become climate-neutral in the near future by using internal combustion engines with synthetic or regenerative fuels.

Airbus, Rolls-Royce, Nestlé and the German Aerospace Center (DLR) are currently investigating the behavior of such fuels in aircraft engines, such as the turbofan engine for the Airbus A 350 passenger aircraft.

The most developed group of sustainable fuels **SAF (*Sustainable Aviation Fuels*) is *HEFA* (hydro-processed esters and fatty acids**), which is composed of *fat waste and waste from the food industry*, among other things. The performance and emissions of aircraft engines powered by HEFA fuel are also being investigated in another current project, ECLIF3 (*Emission and Climate Impact of Alternative Fuels*), in which *Airbus, Rolls-Royce*, manufacturer *SAF, Nestlé* and *the German* Aerospace Center (DLR) are involved.

In this context, the use **of a biokerosene** made from solid raw materials such as grain or algae in aircraft engines was recently tested. SAF also includes so-called **"power-to-liquid"** fuels, which are also produced by synthesizing hydrogen and carbon dioxide. The great advantage of SAF fuels is that they do not require any modifications to the engines, the aircraft itself or the airport infrastructure.

Combustion remains, apart from nuclear fission and fusion, the mother of all chemical reactions with energy release! By recycling carbon dioxide in nature, internal combustion engines with synthetic and renewable fuels do not cause climate change and are almost irreplaceable on land, in the air and at sea.

References for Chapter 2

[1]   https://dw.com/de/afrika-autoindustrie-vda-automarkt

[2]   https://www.energy.gov/eere/fuelcells/hydrogen-production-electrolysis

[3]   www.bundestag.de/resource/blob/190702/2728340e1835bac972eaa07bc4b2e2ca/hochspannungs-gleichstrom-uebertragung-data.pdf

[4]   Stan, C.: Thermodynamics for Mechanical Engineering and Vehicle Construction, 4th edition, Springer Verlag, 2020,
ISBN 978-3-662-61789-2

[5]   Stan, C.: Alternative Propulsion for Automobiles, Springer, 2020,
ISBN 978-3-662-61757-1

[6]   Stan, C.: Alternative Propulsion Systems for Automobiles, Springer International, 2017

Chapter 3

Heat pumps: heat transport against nature needs work

*"**Heat pumps should be the heating system of the future. By 2030, the German government plans to install six million heat pumps. From 2024, we will install 500,000 new heat pumps every year," several government officials, quoted by, said in media, in March 2023. And further:** "Heat pumps should be a climate-friendly heat supply for German households - unlike oil and gas." Admittedly, individual solutions are permitted by federal law – heating network, biomass, hydrogen governors, pellets – but with 65% renewable energies.*

Conclusion: Funding from federal funds, still in 2023, 75%, then 30%. And according to most providers, "air source heat pumps are "significantly cheaper than geothermal and groundwater heat pumps" <u>As a result</u>, at least 500,000 new, essentially <u>winter-sucking heat pumps</u> were to be installed annually to heat houses from 2024 onwards - by 2030 this figure is expected to rise to six million.

Of the three options: a) suck in "cold-air" in winter, push through a little with a fan; b) heat exchangers deep in the earth, which almost always remains warm; c) flowing wastewater, which is always hot, the latter is clearly the most effective: between **the heat** (*as energy supplied to the heating system*) and the **work** on the mostly electric pump (*energy, i.e., electrical energy that is supplied*), the efficiency is almost double between waste water and "cold -air", as sources. On the other hand, in terms of *<u>investment and carbon dioxide emissions</u>*, the cold-air heat pump is no better than a well-tuned gas heating system. You don't need any science for this, you just have to have driven a Trabant in the old GDR. Every driver, whether he had a wife or not, had ladies' stockings on board, which was understandable: the engine was air-cooled, provided with ribs, in anticipation of a minimum of airflow, which came from a fan that listlessly followed its V-belt. Now it often happened that the V-belt broke in front of the defiant fan. After two minutes, the engine was hot, then the pistons stuck firmly into the cylinder bushings. As far as the effect of the flow in a heat transfer, even if the example is reversed, from cold to warm, i.e., not from "warm" to cold. What effect would the heat exchange have had if there had been a strong flow of water with a thousand times the density around the two "Trabbi"s (the nick name for the GDR Trabant car) cylinders? Then the Trabbi would have become a Porsche!

Let's leave the super power of the Porsche engine, the modest power of the Trabbi engine, however the Work is the matter:

The heat pumps, whether with air, earth or wastewater as an exchange medium, basically have the task of "pumping" **energy as heat** from a "cold" to a "warm" source, and this costs, according to the energy balance for the cyclic process in the working medium, a corresponding energy, usually as **work.**

3 Heat pumps: Heat transport against nature requires work

3.1 Natural process of heat flow

Natural processes always have a certain and therefore irreversible direction.

A balancing process of *pressure, voltage, drop height, density (concentration) or heat* always proceeds in the direction of equilibrium. A process in the opposite direction can only occur with the application of an <u>external</u> energy, but this causes a change in the environmental state.

*Example: A flow proceeds naturally, always from **higher** to **lower** **pressure**. Conversely, a flow from the lower to the higher pressure is, of course, also possible, but not as a balancing process, i.e. not in a natural way. This requires **work from the environment**.*

Heat can pass from a *system, subsystem, source* with a comparatively **higher temperature** to an area with a **lower temperature**, but this requires a **work** Figure 3.1.

Definition

Heat is the form of energy that is transferred between two systems *(or between the system and the environment)* of different temperatures as a result of their thermal contact – without any mandatory change in the macroscopic properties of the system, in this case without additional energy exchange, *as long as the sequence of the process is not a balance.*

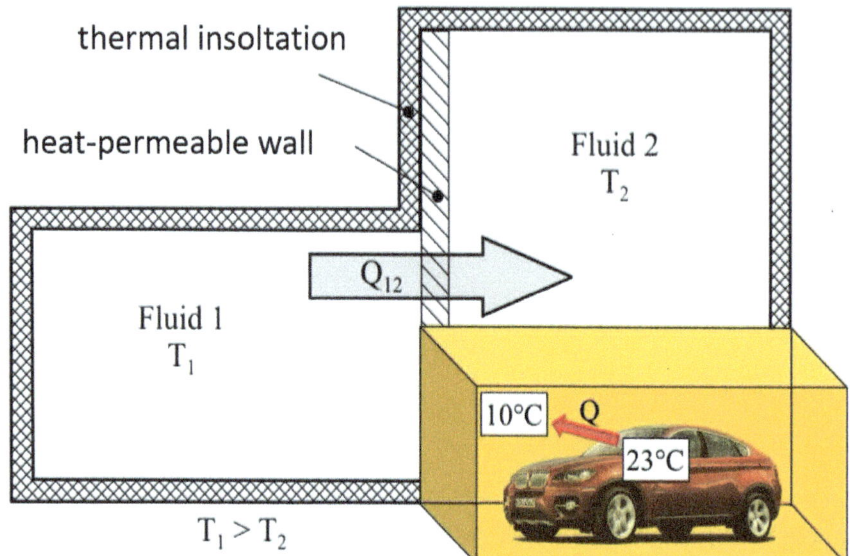

Figure 3.1 Exchange of heat by natural equilibrium between two systems in direct thermal contact

In **refrigeration** machines and heat **pumps**, (*unlike heat **engines**, such as internal combustion engines or gas turbines, transforming a supplied heat into work, as an output*),

work is used to transport a heat flow from a **warm** to a **cold** environment that is not possible as a natural balancing process.

This is the heat absorption at a lower temperature than the heat output from the system (**refrigeration systems**) or the heat output at a higher temperature than the heat input into the system (**heat pumps**). The resulting, **negative heat balance** is possible, but with a corresponding amount of work for the "heat transport".

3.2   Heat pumps and refrigeration systems: balance of heat and work

In connection with the thermodynamic processes in a **heat pump**, an energy balance gives the follows: the system's own energetic variables (*state variables*) **internal energy** and **enthalpy** as well as the energetic variables **heat and work** caused by the process flow (*process* *variables*) in the working fluid between the modules of the machine (compressor, heat exchanger, relief module). [1] Internal energy and Enthalpy are defined in this relationship.

Definition

The **internal energy** (*energy quantity of the system in a state*) – *mostly used for* <u>open</u> <u>systems</u> *(open pipes, with flow of the working medium, i.e.* <u>with</u> *mass exchange) – is the totality of the forms of energy – including the internal energy of the working medium – that have contributed to the system reaching a state. This also includes the "pumping work" (pv, pV) and the kinetic energy of the flow itself.* [1]

Definition

The enthalpy (*energy quantity of the system in a state*) – mostly used for <u>open</u> <u>systems</u> *(open pipes, with flow of the working medium, i.e.,* <u>with</u> *mass exchange)* – is the totality of the forms of energy – including the *internal energy of the working medium* – that have contributed to the system reaching a state. This also includes the "pumping work" (pv, pV) and the kinetic energy of the flow itself. [1]

In addition to the direct energetic quantities, another one, the **entropy**, is often used, which, by multiplying it by the respective temperature of the working medium, represents a heat that can also be graphically drawn as a surface. Due to the direction of the thermodynamic process, it is possible to see in the diagram whether such heat is supplied or dissipated, which appears to be very clear for a heat balance.

Definition

Entropy (*energy quantity of the system in a given state*) is a <u>numerical measure of the irreversibility of a thermodynamic process</u> and as such can be calculated with equations

---

It would be so simple if one would leave the entropy as an "operator" for the degree of irreversibility of a process and calculate it pragmatically, with a respective equation, instead of dragging it into "entropy export" for processes without obvious heat or energy exchange, which would lead to the "cold death" or the "heat death" of the universe! [1]

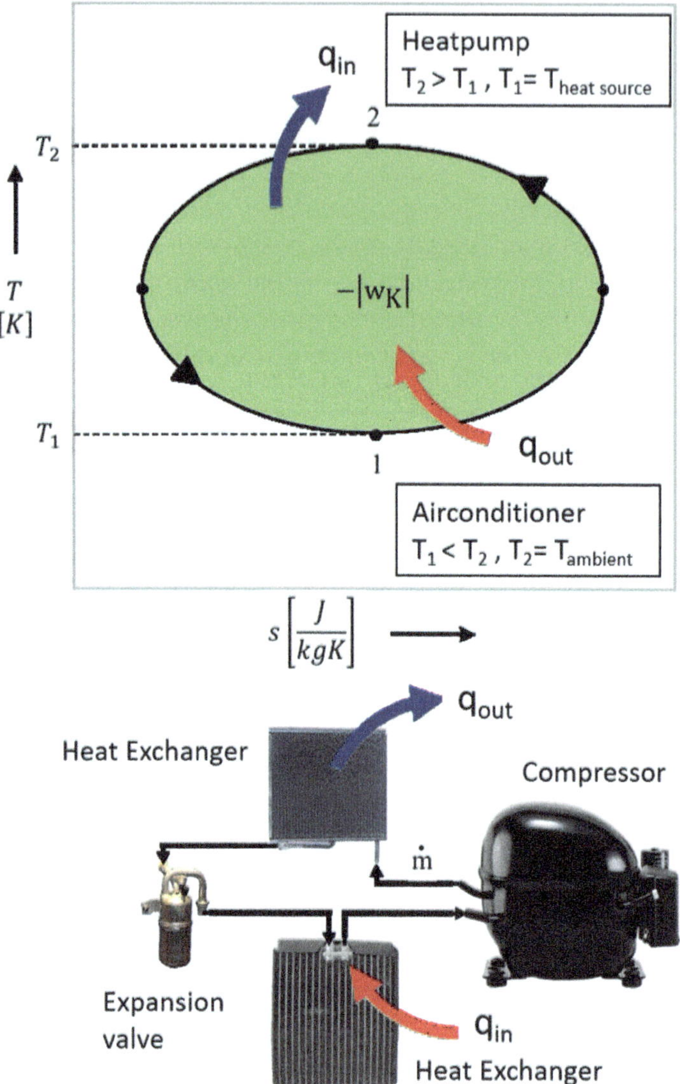

Figure 3.2 Refrigeration systems and heat pumps – the name depends on the use side of the heat flow process

According to the principle of heat flow from a warm to a cold environment with the help of work from the system environment, *(not possible as a natural balancing process)*, both **refrigeration systems** and **heat pumps** works similarly. The only difference between air conditioners and heat pumps is which side of the heat flow – heat input in a heat exchanger, or <u>heat output</u> via heat exchanger is used. Figure 3.2 shows schematically the basic process for such applications.

Figure 3.3 shows an example of a liquid working fluid, in this case water (*which is not suitable for air conditioning systems and heat pumps due to freezing temperatures and other evaporation properties*).

The <u>energy balance</u> between heat and <u>work</u> is derived from the respective changes in state (*compression, heat dissipation in rooms, relief, heat supply from the environment*):

$H = U + pV + \dfrac{mc^2}{\underbrace{2}}$ The following applies (Figure 3.4 for R134a):

neglected in this case

<u>Energy-Bilance</u>

Compression: $q_{12} - w_{12} = h_2 - h_1$; with $h_2 > h_1$
(1, 2)   (no heat exchange)

Heat Transfer by heat exchanger to rooms:
(2, 3) $\underbrace{q_{23}} - w_{23} = h_3 - h_2$; with $h_3 < h_2$

                                     (q_{in})
                                     (no work in the heat exchanger)

Expansion: $q_{34} - w_{34} = h_4 - h_3$; ideally $h_4 = h_3$
(3, 4)   (no heat exchange)

Heat transfer to heat exchanger: $\underbrace{q_{41}} - w_{41} = h_1 - h_4$; with

$h_1 > h_4$
(4, 1) (q_{in}) (no work in the heat exchanger)

Sum: $w_{heatpump} = |q_{ab} - q_{zu}| = |h_2 - h_3| - (h_1 - h_4)$

Coefficient Of Performance (COP):

$$cop = \frac{q_{out}}{w_{heatpump}} = \frac{|h_2 - h_3|}{h_1 - h_4}$$

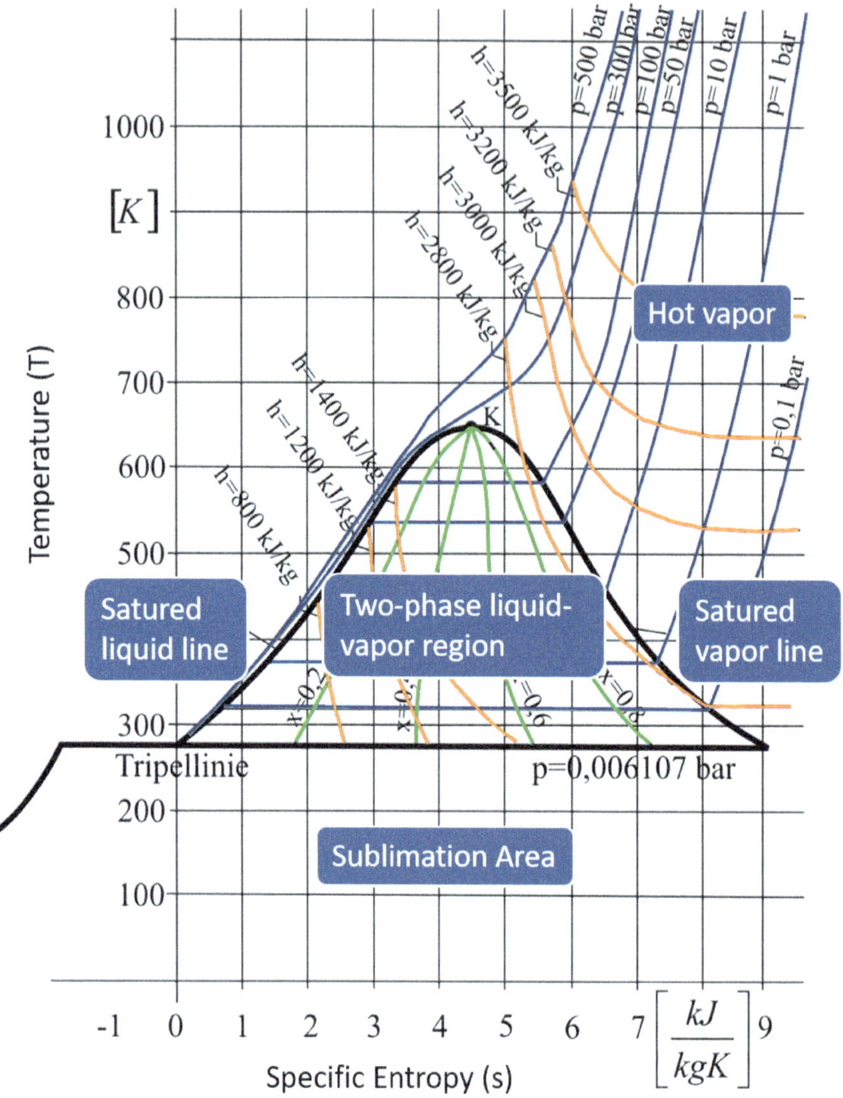

Figure 3.3 T,s - Diagram for water vapor, a steam-capable working device, in basis on the representations for enthalpy, pressure and temperature [1]

In the wet vapor region in which the working fluid is partly liquid, partly gaseous, partly vaporous (*represented by the factor x, as a percentage of the Steam – x=0.6 means 60% steam) remains constant (e.g., temperature and pressure)* for any change in state, according to Gibbs' phase control, [1] for not one, but two state variables. After the upper boundary curve, as in the liquid phase, they again have different courses to the left of the lower boundary curve.

For working fluids suitable for both refrigeration systems and heat pumps, one of the usual forms of representation (log pressure p [megapascals) / specific enthalpy (kJ/kg working fluid) is used, as shown in Figure 3.4.

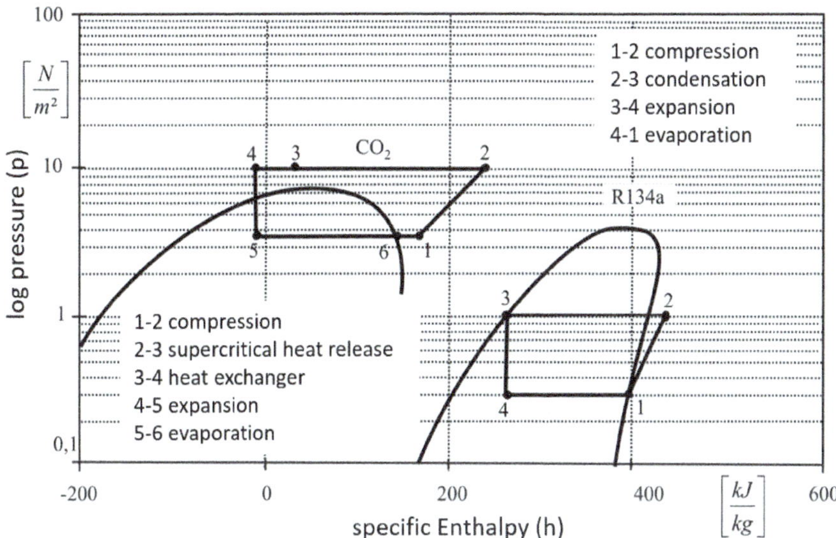

Figure 3.4 Working diagram for two working fluids, **CO₂** and **R134a,** which are suitable as fluids in refrigeration systems and in heat pumps, in use, in a common diagram form: pressure logarithm/ specific enthalpy [1]

3.3   Heat pumps and refrigeration systems: thermodynamic processes of working fluids

In a heat pump, the room to be heated is located after the compressor outlet (2), as shown in Figure 3.5 (T,s diagram). For example, the working fluid can reach a temperature of around 50 – 60 [°C] after compression.

Heat transfer through the heat exchanger lowers the temperature (2 to 3), but the pressure remains constant. In the condenser, heat is expanded at the same pressure, while the temperature also remains "the same", which can be misleading: the vapor content drops from x=1 to x=0, *(according to* Figure 3.3*)* in which the working medium becomes liquid. A further heat expansion, across the lower limit curve (4 to 5), then clearly shows that the temperature is falling.

In the throttle (5 to 6), the working medium expands, without further energy exchange with the environment, using its own internal energy to evaporate, which can be seen in the increase in entropy (5 to 6). On the change of state in the next, flow-permeable heat exchanger, the pressure remains constant (6 to 7),

Note: in the *first part of the flow at constant pressure (2 to 3) in the upper heat exchanger (condenser), the property of the liquid substance is also used to give off heat during liquefaction, which reduces the temperature all the more. This is analogous to the process on the evaporation side, where a corresponding enthalpy is extracted from the working fluid itself, which results in its further cooling in the evaporator.*

The temperature at the end of the discharge is therefore lower than the temperature of the medium that <u>flows</u> into the heat exchanger (evaporator) (6 to 7), which can be either *air, warm river water or even warmer waste water or <u>surrounding</u> warm earth).*

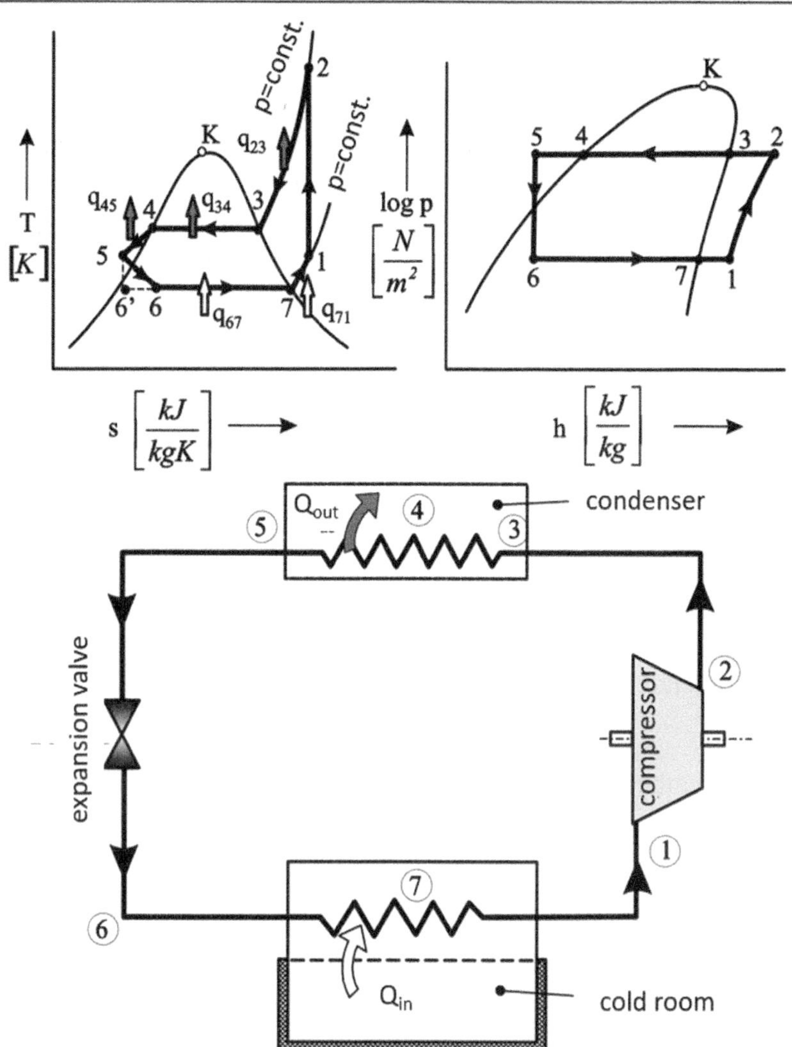

Figure 3.5 Thermodynamic cycle in the working fluid in two forms (T, S) and (log p, h) – diagram and schematic representation of the functional modules of a heat pump

The temperature of the working fluid thus reaches the value at the compressor inlet and the thermodynamic cycle can be started again from the same state. The change of state (7 to 1) takes place still in the heat exchanger, but the working medium is already in the complete gas phase, as superheated steam, according to the upper limit curve.

The energy balance in such a thermodynamic cycle can be deduced in the diagram log p/h in Figure 3.5 as the distance on the abscissa between the specific enthalpies in heat dissipation ($h_5 - h2$) and the work by the compressor ($h_2 - h_1$).

Definition

The **Coefficient Of Performance COP of the heat pump** is the ratio between the enthalpy difference of the working fluid in the heat dissipation and the enthalpy difference (work) in the compression.

This ratio can be "read" from an h,s or log p,s diagram of the work equipment in the heat pump's thermodynamic cycle as a distance comparison. The enthalpy difference (distance in heat dissipation $h_{3-h}2$) in relation to the enthalpy difference in compression, (distance $h2-h_1$)

The efficiency is generally given as COP (Coefficient Of Performance).

In mild weather, the coefficient of performance (COP) can be around 4, [2] while an air source heat pump can still achieve a COP of 3 at temperatures below about $-7\ °C$ (19 °F) [3].

3.4   Heat pumps: configuration examples

Figure 3.6 shows the embodiment example for a common domestic heat pump.

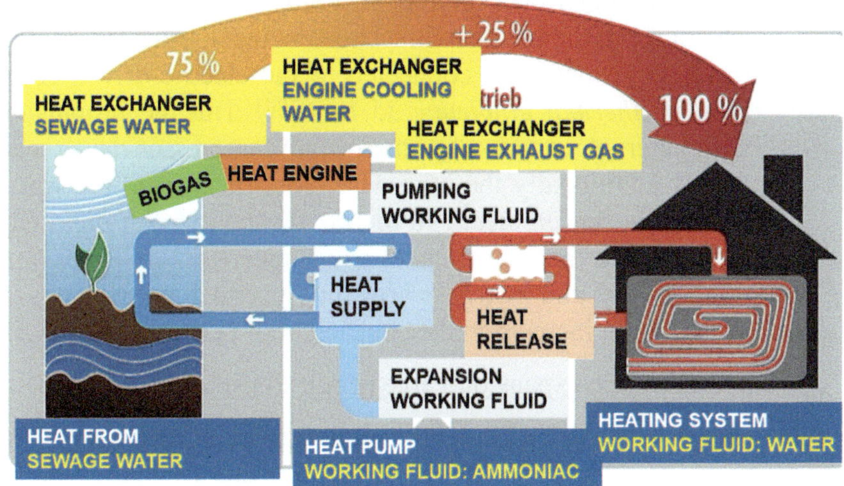

Figure 3.6 Design example of a heat pump

The central module shows the basic heat pump:

- *Working fluid* in a closed circuit, in this case the currently commonly used R134a (tetrafluoromethane), otherwise mixtures of difluoromethane and pentafluoro methane [4] or other combinations of hydrocarbons, such as, for example, pentane [5] or ammonia.

- *Heat exchanger* (evaporator) on the side of the heat source (earth, wastewater, water flow, *ambient air*) with liquid working fluid, with convective heat transfer to the flow of the working fluid (R134a) in the heat pump.

- *Heat exchanger* (condenser) on the consumer side (*underfloor heating pipes, radiators on the wall*) with liquid working fluid, with convective heat transfer from the flow of the work fluid (R134a) in the heat pump.

- *Compressors* generally designed for small and medium-sized heat pumps as scroll compressors driven by an electric motor [6].

  For higher outputs, *piston engines* are used as compressor drivers, more climate friendly with combustion of *biogas*, green *hydrogen* or a mixture of both. There appears a main advantage of the variant with piston engine: the electric energy for electric motors is connected to the electric-driven compressor *(thermal efficiency 35 to 45%)*. But the propulsion with internal combustion engine as compressor-driver can supply additionally the otherwise "lost" heat from the water for *engine cooling heat* (90 to 100°C) and *exhaust pipes* (up to 700°C) to the heat (intake) exchanger. This reduces compressor work and increases heat supply, which corresponds to a significant increase in the coefficient of performance (COP)

- Thermostatic expansion valve (throttle) [7]

When using R 134a (tetrafluoro ethane) as a working fluid, the pressure range is 2.93 bar to 13.18 bar, or the temperature range 50°C – 0°C [8].

At temperatures higher than 50°C on the heating side, the work equipment is often ***"reheated" with the help of an electric heating element! Such a solution is equivalent to gasoline heating, which is also often used, in a car with an electric motor and battery!***

After all, the energy efficiency of such a heat pump is only shown by the coefficient of performance (COP): *how much heat the system has emitted for heating, and how much energy it has used, as electricity or as fuel.*

Various data sheets give figures between 2.7 and 4.8, which depend on the system itself, but also on the heat extraction.

Seasonal coefficient of performance factor of the heat pump

However, a more objective measure of the efficiency of a heat pump is the annual performance factor.

> Definition
>
> *The seasonal coefficient of performance: Heat flow output to the heated room in relation to the compressor output, both integrated over time in one year.*

For a single-family house with a heating energy requirement of 150 kWh/ (square meter * year) [9] the following annual performance factors are given for various forms of heat transfer from the heating medium to the heat exchanger (evaporator) [10] :

- 2,5 for heat from the ambient air

- 3,6 for heat from the ground

- 5 for heat from wastewater or river water.

The numbers speak for themselves: the continuously flowing liquid wastewater, at temperatures (on average in winter) of 10°C to 12°C, provides a *convection* that efficiently supports the heat transfer to the heat exchanger. On the other hand, the air as a gas, with low density, even with a flow forced by the fan, has a much to low efficiency in heat transfer.

The efficiency of the heat pump ambient air over the year – remains a half of that achieved with wastewater!

3.5   Heat sources for a heat pump

A **ground-source heat pump** as shown in Figure 3.7 extracts the soil or groundwater, remaining at a relatively constant temperature throughout the year, around 8°C-12°C at depths of 10 to 100 meters. Geothermal energy can be obtained in two ways: either by *means of geothermal collectors*, which are installed close to the surface, on areas 1.5 to 2 times larger than the area to be heated, or by *geothermal tubes*, which penetrate up to 100 meters deep into the earth. *A brine* of frost-proof liquid circulates through geothermal collectors and geothermal probes. A well-maintained geothermal heat pump usually has a coefficient of performance of 4.0 at the beginning of the heating season and a seasonal coefficient of performance of around 3.0 when heat is extracted from the ground.

Air-source heat pumps are used to bring heat into the heat exchanger through forced convection by means of a fan. These air heat exchangers can also be operated in a cooling mode, in which they extract heat from the refrigerator via internal heat exchanger and release it into the ambient air via the external heat exchanger. Air-source heat pumps are relatively easy and inexpensive to install, making them the most widely used type of heat pump. In mild weather, the coefficient of performance (COP) can be around 4.

While older air source heat pumps performed relatively poorly at low temperatures and were better suited to warm climates, newer models with variable speed compressors remain efficient even at low temperatures, resulting in widespread adoption and cost savings.

In some applications, a heat pump can usually be used to reverse the direction of the heat *flow* by means of a diverter valve (*reversible heat pump*). The diverter valve changes the direction of the refrigerant by the circuit and therefore the heat pump can deliver either *heat* or *cold* to a refrigerator or to a room.

The two heat exchangers, the condenser and the evaporator, which need to reverse their functions, are optimized to function adequately in both modes. Therefore, the seasonal performance factor of a *reversible heat pump* is usually slightly lower than that of two separately optimized machines.

3.6   Heat pumps: areas of application

Heat pumps are also used as a heat supplier for district heating.

In Europe, heat pumps account for only 1% of the heat supply in district heating networks.

However, several European countries have set themselves the goal of reducing anthropogenic emissions from heating: Possible heat sources for such applications are *wastewater, seawater, lake and river water, industrial waste heat, ash streams, flue gas*. Since the

1980s, more than 1500 MW of large-scale heat pumps have been installed in Europe, of which around 1000 MW were in use in Sweden in 2017.

Industrial Heating

There is great potential to reduce energy consumption and associated greenhouse gas emissions in industry through the use of industrial heat pumps. An international cooperation project (2015) merged a total of 39 examples of R&D projects and 115 case studies worldwide [11]. The studies shows that short payback periods of less than 2 years are possible, while at the same time achieving a high reduction in CO 2 emissions (in some cases more than 50%) [12],[13]. Industrial heat pumps can heat up to 200 °C and meet the heating needs of many light industries [14], [15]. In Europe alone, 15 GW of heat pumps could be installed in 3,000 plants in the *paper, food and chemical* industries [16].

A typical heat pump has COP values between 3 and 5 (outside temperature 10 °C / internal temperature 20 °C). The ground is a practically constant source of temperature, which is why ground-source heat pumps are not subject to any fluctuations and are therefore an energy-efficient form [17].

3.7   Use of heat pumps in the world

Australia: The use of *reversible heat pumps is being investigated* in the food processing industry, breweries, pet food manufacturers and other industrial energy consumers. *Process heat* accounts for the largest share of on-site energy consumption in Australian manufacturing, with lower-temperature operations such as food production particularly well-suited to the transition to renewables.

To help producers understand how they could benefit from the transition, the Australian Renewable Energy Agency (ARENA) provided funding to the Australian Alliance for Energy Productivity (A2EP) to conduct pre-feasibility studies at a number of sites across Australia, with the most promising sites moving to full feasibility studies [18].

Canada: In 2022, the Canada Greener Homes Grant [19] provides up to $5000 for upgrades (including certain heat pumps) and $600 for energy efficiency assessments.

After the *Inflation Reduction Act* was passed by the United States Congress and signed into law by President Joe Biden on August 16, 2022, *the High-efficiency Electric Home Rebate Program* was created to provide grants to state energy boards and Native American tribes to introduce statewide rebates for high-efficiency electrical homes. As of now, American households are eligible for a tax credit to cover the cost of buying and installing a heat pump of up to $2,000. As of 2023, low- and middle-income households will be eligible for a heat pump discount of up to $8,000 [20].

USA: In 2022, more heat pumps were sold than natural gas heaters in the United States [21].

Some U.S. states and municipalities have already offered incentives for air source heat pumps:

California

In 2022, the *California Public Utilities Commission* added an additional $40 million to the existing $44.7 million budget of the *Self-Generation Incentive Program* (SGIP) *Heat Pump Water Heater* (HPWH) program, where single-family home customers can receive an incentive of up to $3,800 to install an HPWH, from the Auction proceeds allocated for 2023 gas cap-and-trade certificates. Half of the incentive funds are reserved for low-income utility customers, who are eligible for a maximum incentive of $4,885 [22].

Maine

The Efficiency Maine Trust offers residential heat pump rebates of up to $1,200, as well as low- and middle-income Maine heat pump rebates of $2,000 for their first eligible heat pump and up to $400 for a second eligible heat pump [23], [24].

Massachusetts

Mass Save, a joint initiative between Massachusetts' natural gas and electric utilities and energy efficiency service providers, offers a discount of up to $10,000 for *air-source heat pumps,* covering the purchase price of the heat pump and installation costs [25].

Minnesota

Minnesota Power is offering an *air-source heat pump* discount of up to $1,200 if the pump is purchased and installed by a participating *Minnesota Power* contractor [26].

Europe

The heat from *wastewater from residential and industrial buildings or from cooling water* from heat engines and other equipment or processes can already be transferred from a temperature of 10°C – 20°C to a working fluid in a closed circuit through heat exchangers, thereby transferring temperatures of 70 – 80 °C via heat exchangers to a heating system. Appropriate work is required for working fluid.

If a wastewater flow passes through a pipe with a diameter of one meter, is wrapped by a spiral heat exchange pipe over a length of 70 meters and thus emits only 4 °C, a heat flow of 33 kilowatts is already generated [27]. This heat flow is transferred to the working fluid in the heat exchange tube.

The damped working fluid can then be brought to a higher pressure with a compressor. this also increases its temperature, for example up to 70 °C.

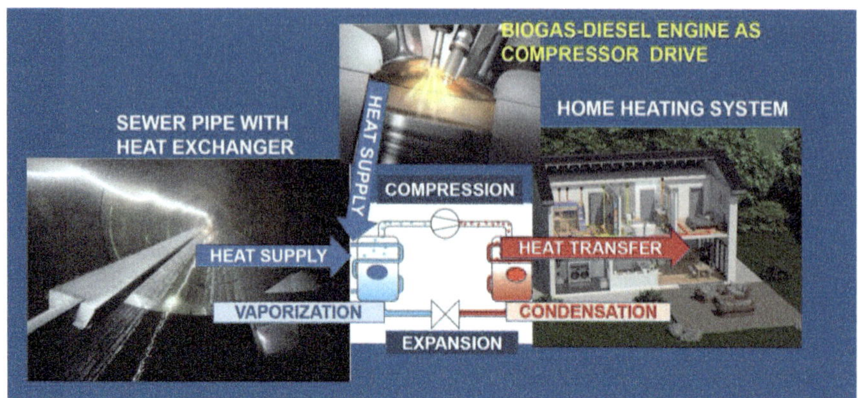

Figure 3.7 Functional diagram of a heat pump using waste water as the main heat source and cooling water and exhaust heat from the driving heat engine as auxiliary heat sources

An electric motor for the compressor-actuating internal combustion engine has an efficiency of 90 – 95%, but the electrical energy is generally supplied to it by a power plant. Depending on the energy source used – *coal, natural gas, crude oil or nuclear energy* – this has an efficiency of 25% to 50%. As a result, the overall efficiency remains below 47% when using a compressor with an electric motor.

Such efficiency is usually also achieved by a newer generation of stationary diesel engines. Of course, the entire energy chain from oil production and refining to transport to the point of use should also be considered in this case.

A good example of a heat pump in the living area is the system in a block of 78 apartments in Berlin-Karlshorst, where 80% of the heat is covered with a heat pump. Carbon dioxide emissions are thus halved compared to a conventional gas heating system. Such installations can be widely used when pipes with an integrated heat exchanger are laid during the renewal of the sewage pipes of a local or residential area, as shown in Figure 3.8.

Berlin's sewer network alone consists of channels for *wastewater, rainwater and mixed water* with a total length of 9,400 kilometers! And this mixture is always warm, in summer as well as in the depths of winter. This is a gigantic heat that has so far been almost completely lost by transferring it in the many clarification stages of the atmosphere.

Figure 3.8 Sewage pipes with integrated heat exchanger

In the town of Kalundbord, Denmark, there are three heat pumps with 3.3 megawatts each, which use *wastewater* from wastewater treatment plants at temperatures of 15-30 °C, from which so much heat is extracted that the temperature drops by 5 °C via the heat exchanger. With each of the 3 plants in which the working fluid is *ammonia*, so much heat is transported that the water temperature in the heating circuit rises to 80 °C. The 3 compressors are each driven by four piston engines. This supply of heat and hot water benefits 4400 inhabitants.

The most spectacular example of an efficient heat pump, however, is the *Hotel Carlson in St. Moritz, Switzerland*. The hotel has numerous suites, indoor and outdoor pools, saunas and steam rooms, for which the water must be circulated and heated. The corresponding energy consumption is 800,000 kilowatt hours per year. In a classic heating system with a combustion chamber under a boiler, 80,000 liters of heavy fuel oil would be burned for this purpose. The more advantageous solution in terms of energy consumption and especially for the air in St. Moritz was to install a heat pump during the reconstruction of the hotel in 2007, which is supplied with the heat of the collected flows of wastewater from baths and pools.

These wastewater flows are collected in a concrete storage tank with a capacity of 25 cubic meters and release as much heat through a heat exchanger until their temperature drops by 3°C before being discharged into the city's wastewater system. In this application, the heat pump is driven by an electric motor in order to completely avoid noise and pollution around the hotel. With this system, the fresh water is brought to a temperature of 60 °C.

Such solutions should indeed be used on a broad scale, especially in building heating and hot water preparation, because this is the category with the largest share of primary energy consumption on a global scale – ahead of industry, construction, cement production or transport.

References for Chapter 3

[1] Stan, C.: Thermodynamics for Mechanical Engineering and Vehicle Construction, 4th edition, Springer Verlag, 2020,
ISBN 978-3-662-61789-2

[2] "Wärmepumpen mit Prüf- / Effizienznachweis (heat pumps with efficiency validation)" (in German). BAFA (Federal Office for Economic Affairs and Export Control in Germany). Retrieved 2022-02-20

[3] https://bafa.de/wärmepumpen

[4] https://infraserv.com/leistungen

[5] Wu, Di (2021). "Vapor compression heat pumps with pure Low-GWP refrigerants". *Renewable and Sustainable Energy Reviews*. 138:110571.
doi:10.1016/j.rser.2020.110571

[6] Walter Grassi: *Heat Pumps, Fundamentals and Application*, Springer Verlag 2018, ISBN 978-3-319-62199-9

[7] https://icold.world/de/know-how/thermostatische-expansions-ventile-tev-fur-kalteanlagen

[8] Walter Grassi: *Heat Pumps: Fundamentals and Application, Green Energy and Technology*, Springer International, Cham (Schweiz) 2018, ISBN 978-3-319-62198-2

[9] https://heizungsfinder.de/waermepumpe/wirtschaftlich-keit/jahresarbeitszahl

[10] https://ost.ch/de/forschung

[11] IEA HPT TCP Annex 35 Archived 2018-09-21 at the Wayback Machine

[12] IEA HPT TCP Annex 35 Publications Archived 2018-09-21 at the Wayback Machine

[13] IEA HPT TCP Annex 25 Summary Archived 2018-09-21 at the Wayback Machine

[14] "Norwegian Researchers Develop World's Hottest Heat Pump". Ammonia21. 2021-08-05. Retrieved 2022-06-07

[15] "Heat pumps are key to helping industry turn electric". World Business Council for Sustainable Development (WBCSD). Retrieved 2022-10-04

[16] Technology Report: The Future of Heat Pumps. International Energy Agency (Report). November 2022. License: CC BY 4.0

[17] "Heating and cooling with a heat pump: Efficiency terminology". *Natural Resources Canada. 8 September 2022. Retrieved 3 April 2023*

[18] "Electrifying industrial processes with heat pumps". 22 March 2022. Retrieved 2022-08-09

[19] "Canada Greener Homes Grant". 17 March 2021. Retrieved 2022-01-17

[20] Elena Shao. "H.R.5376 - Inflation Reduction Act of 2022". Congress.gov. U.S. Congress. Retrieved 17 November 2022

[21] "As Heat Pumps Go Mainstream, a Big Question: Can They Handle Real Cold?". The New York Times. February 22, 2023

[22] "CPUC Provides Additional Incentives and Framework for Electric Heat Pump Water Heater Program". cpuc.ca.gov. California Public Utilities Commission. Retrieved 16 November 2022

[23] "Residential Heat Pump Rebates". Efficiency Maine. The Efficiency Maine Trust. Retrieved 16 November 2022

[24] "Heat Pump Rebates for Low and Moderate Income Main-
      ers". Efficiency Maine. The Efficiency Maine Trust.
      Retrieved 16 November 2022

[25] "Air Source Heat Pump Rebates". Mass Save. Retrieved 17
      November 2022

[26] "ASHP Rebates". Minnesota Power. Retrieved 17 Novem-
      ber 2022

[27] Stan, C: Future Fire Forms, Springer 2020

Chapter 4

Photovoltaics for electrical energy and for climate-friendly fuels

Photovoltaic solar modules are promising solutions, with **43** to **134 W per square meter** of panel area (amorphous or monocrystalline silicon), but in practice the value levels of around **100 W/m²**. There is even talk of **200 W/m²**, however, under laboratory conditions and with vertical solar radiation. The electrical power [W] per square meter [m²] is an efficient indicator of how large the panel area must be for a desired performance.

Nevertheless, from the start ramp, it immediately goes to all means of mobility, if not satellite, then car or boat.

In Europe, the scenario of **an electric** car with 20 to *30 kWh/100 km, at 20 km a day,* with *10-15 m² of solar panels on the roof* of the house (if the sun is shining), *for the 2 to 3 kWh* that the car then consumes, seemed to be the   assumed standard values, so satisfactory that the authorities opened the funding pots at lightning speed. This would even apply to almost all possible surfaces of cars, for example to Toyota, and Hyundai too. However, they actually got an output of only 200 watts for the total of five square meters of car space. So, it's better to **go on boats**, precisely because of their surface area on the wider water: *400 watts maximum power, 6 to 10 kWh battery, 10 kW drive motor*. On Lake Garda, in Limone/Italy, our students of the West Saxon University of Zwickau/Saxony/ Germany had very good experiences with such a boat.

However, the situation is different on houses: 2880 kWh/month were measured and reported in northern Germany in summer, 1220 kWh/month in winter, with a more specification: 625 kWh/month in August but then 90 kWh/month in November.

This is far too little for a house in northern Germany with a yearly sum of 4000 kWh, and only a tenth of the expected energy. Incidentally, this also corresponds to the contribution of solar energy to primary energy on a global scale.

The solution would be simple, as long as there is space for it: you just have to increase the area! Examples can be found from Arizona (*Agua Caliente*) to the *Sahara, Kalahari/* (Namibia, Angola).

The thermodynamics of solar radiation offers an interesting initial point for this: *The* Energy flux of solar radiation is, as an integral of the respective energy density over the wavelength of light, referred to as **the solar constant** and fixed at **1361 W/m²**, after numerous scientific measurements. In practice, however, very little is "harvested": on 2700 hectares in China, it is **only 3 W/m²** in the whole of 2018.

Physics justifies this in terms of the material: The silicon cells in the solar panels have phosphorus atoms on the upper side and boron atoms on the lower side. Even if they received enough light radiation, they would eventually be saturated in terms of reaction. Alternative material combinations, from cadmium telluride to copper-zinc-tin sulfide, dye-sensitized solar cells (DSC) or organic photovoltaics (OPV) do not bring any significant improvements. In addition, it has been proven that *rain, hail, snow load, heat and cold cycles* impair their actual, area-related energy flux density by up to 30%.

In the world, however, *which is also good and right as a climate-friendly measure,* large-scale and small, individual photovoltaic systems are being installed. But despite the "bombastic" promises of their proponents, including output figures in terawatt hours, they provide only 3.8% of the primary energy that can be produced with coal and gas on a global scale. Above ground and latitude, the radiation efficiency of all photovoltaic systems worldwide is 13.7%.

Instead of trying to sow solar panels to the point where it is no longer possible to sow over houses, roads and parking lots, car roofs and tomatoes, in the Black Forest or Lapland, it would be more advisable to build large solar plants where the sun shines strongly, for a long time: *Australia, Kenya, Egypt, Nevada, Colombia, Guatemala.* On the other hand, small photovoltaic systems in individual, modest dwellings, of which there are millions and millions in the world, *in Africa, Asia, Australia or even in Europe* are a step towards true civilization.

The decisive potential of photovoltaics, however, is the production of hydrogen, as an easily and relatively inexpensively transportable energy carrier for machines in Europe and all over the world. Hydrogen as a fuel for industry and heating, hydrogen as a fuel for "combustion engines" and fuel cells in automobiles, hydrogen encapsulated in ammonia, or as a basis for the formation of hydrogenated vegetable oils (HVO-hydrotreated vegetable oils), made from plant residues but also from frying fat from French fries. Electrolytically generated hydrogen by means of photovoltaics, packaged in ammonia and vegetable oils, is logistically far cheaper than storing photovoltaic electrical energy in batteries or transporting it through heavy cables thousands of kilometers long to Europe and America.

4 Photovoltaics for electrical energy and for climate-friendly fuels

4.1 Application of photovoltaics in power generation

Examples of photovoltaic applications in urban and water mobility

According to currently published opinions [1], *photovoltaics* and *wind energy* are the backbones of general energy consumption. According to the definitions in Chapter 1, this refers to final energy, and in particular *electrical energy,* as a secondary form of energy [1], [4].

A classic scenario is the *"photovoltaic sunflower",* as shown in Figure 4.1 and Figure 4.2, which was put into operation in 1997 by our West Saxon University of Applied Sciences in Zwickau (Saxony), Germany.

It consists of 12.7 m² of adjustable solar cells made *of monocrystalline* silicon and of additional 10 m² of non-adjustable solar cells made of amorphous silicon.

© The Author(s), under exclusive license to
Springer-Verlag GmbH, DE, part of Springer Nature 2024
C. Stan, *Energy Scenarios for the Future,*
https://doi.org/10.1007/978-3-662-69687-3_4

This solar system was mainly used as a power supply for the university's own electric *VW Golf*, which supplies the data centers spread throughout the university campus.

Figure 4.1 „Photovoltaic sunflower", solar system at the West Saxon University of Zwickau (Saxony), Germany, in operation since 1997 (*Source: West Saxon University*)

According to the available data [5] the monocrystalline solar modules reach **134 W peak/m²** under ideal standard conditions, while the amorphous modules reach **43.2 W peak/m²**. *W peak symbolizes the maximum achievable power that appears in full sunlight. This is usually also given as* **nominal power** *(target), as the ideal limit.*

The currently available electrical power (is), which could be seen from afar, in passing, for years, every day on the large display of the Zwickau solar system, in bright sunshine and in clouds, whether Easter or Christmas, was 1000 [W] to a maximum of 1200 [W] (is) – this is always less than 100 W/m² for the mentioned area of 12.7 m². Very often, however, the electrical power (is), even during the day, in good light, was zero!

Recently, the Helmholtz-Centrum Berlin/Germany and the Karlsruhe Institute of Technology/Germany have reported on state-of-the-art photovoltaic modules whose efficiencies (deducted) are around **200 W/m²**. [6] This value can be achieved under optimal, precisely defined conditions in the laboratory, with maximum perpendicular light radiation.

In order to compare solar cells with each other, especially in terms of materials or layers used, the **efficiency of the solar cell** is often defined as the electrical energy current it generates to the energy flow of incident solar radiation. About 20 years ago, the average efficiency of a solar cell was 20%, currently it is 47%. [7]

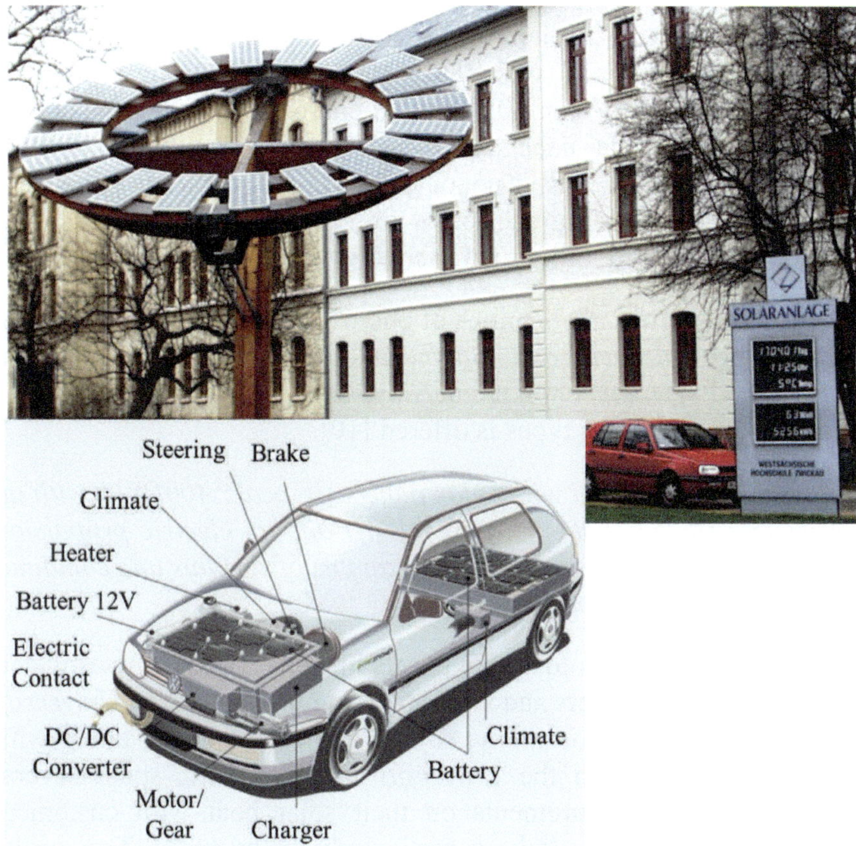

Figure **4.2** Power supply for electric vehicles for journeys between the data centers at the West Saxon University of Zwickau/Saxony, Germany

An individual configuration of photovoltaic station-battery-electric car, as it is currently being discussed in Europe because of state subsidies, seems to be satisfactory:

According to practical data, such a standard compact to mid-range electric car with batteries consumes **20 to 30 [kWh/100 km]** [8], [9].

Driving less than 20 kilometers a day, the battery will be charged with the necessary 2 to 3 [**kWh] per day, with an area** of 12 to 15 [m²] *(just for this task!),* in about two hours, during which the sun shines very brightly *(if it is shining)* solar panels on cars *(vertical surface projection on a car to the sun's rays, always below 5 [m²]),* the charging power with electric energy remains below the requirements, which is necessary in many projects, for example in Toyota Prius Plug-In with solar roof (with 180 watts of solar charging power) <u>*or in Hyundai Ioniq 5*</u> (with 205 watts of charging power) was shown. The areas of the solar panels are therefore to become much larger as shown in Figure 4.3. This is not only provided by the roofs of houses, but also by the water surfaces on or around the boats, as long as they do not interfere with other road users Figure 4.4.

Such advantageous applications of photovoltaic panels are already in use: between *Bregenz and Lindau* or on *Lake Maggiore,* between *Ascona* and *Verbania* or on the *Berlin lakes,* where a fleet of large and small boats of various types is offered [10].

Such a boat, equipped with solar panels (W peak: 400[W]), with an electricity storage battery (6-10 [kWh]) and an electric propulsion motor (10 [kW]) represents a good example of variants and combinations of such utilization.

The application of photovoltaic panels in boats was also investigated by a team of researchers and students from our *West Saxon University of Zwickau/Saxony, Germany,* several years ago. They, along with their colleagues from the *University of Pisa, Italy,* spent several months taking measurements on their solar boat with customers aboard Lake Garda in Italy, a particularly sunny place. The results were excellent for this application: when the sun is burning, the solar panels provide a welcome shade, and the boat moves slowly, powered by the little but sufficient power [kW] of the electric propulsion motor, without noise and without pollution. The solar roof of the boat provides a very suitable parasol in the sun anyway, a welcome umbrella in the rain, even if there is no sun shining when it rains to activate the solar panels. When it rains, however, you do not go out to sea in a boat, except in emergencies, for which the storage battery is available.

Examples of photovoltaic application for power generation in buildings

Due to the variety in terms of *building size, number of people, energy requirements and* targeted application, it would be almost impossible at this point to quantify individual required energies and, on the other hand, the available areas for solar modules in a clear framework. The example therefore assumes an already derived energy demand [Ws or kWh]:

A photovoltaic house system in northern Germany, intended for a maximum (target) power **W peak** of 5 [kW], which can be derived as energy over the time of the year [s or h], delivers a total energy of (real) 2880 [kWh] in the summer months (April to September), *but in winter (*November to March) only (real) *1220 [kWh].*

The difference between the individual months shows this dependence on weather and season even more clearly: in August it is **625 [kWh],** in January **135**
[kWh], but only **90 [kWh]** in November! For the year as a whole, the energy to (real) **4000 [kWh].** However, at the maximum planned (target) power (W peak) of **5 [kW], an energy of (target)** 43200 [kWh] **would theoretically be expected in the 8760 hours** of a year – not even a tenth of that was achieved!

In order to get from (actual*)* 4000 {kWh] to (target) 43200 [kWh], *the area should be* increased tenfold, as far as this is possible: This is already practiced in the Arizona desert or on the seas.

On the other hand, the roofs of the individual houses (50 to a maximum of 150 [m²]) and their (*preferably)* horizontal projection surfaces are clearly limited.

Figure 4.3 Surfaces with solar panels on earth

Figure 4.4 Surfaces with solar panels on water

4.2   Thermodynamic Fundamentals of Solar Radiation

In idealized scenarios, some proponents [11], calculate how much _energy the sun could theoretically provide to the earth annually through radiation_ and then divide this energy among the world energy demand of a given year. What comes out of this is fascinating and "virtually" reassuring at first: the sun provides tens of thousands of times the energy we need!

Why don't we use them in clever investments?

For an objective assessment, a systematic analysis of such information is recommended, considering the thermodynamic basis of solar radiation, as follows:

First of all, the area-related energy **flux density (radiation intensity of the sun)** [W/m³] Figure 4.5 over the immense areas of the earth and the oceans becomes the **energy flow [W] and over time becomes** energy [Ws or kWs, kWh or J, kJ, MJ]**, which is converted into internal** energy via heat transfer [**Ws or kWs, kWh or J, kJ, MJ**, but also, for plants and humans - **kcal**], by plants, humans and water, which can be seen as a **temperature** increase [K]. As a result, the energy flux density [W/m²] (as defined below) is significantly reduced [1].

The average, area-related **energy flux of solar radiation Q [W/m²]** [1]), at the boundary of the Earth's atmosphere, as an integral of the respective Energy **flux density** [W/m ³] over the corresponding **wavelength of light** [1/m] or proportional to its reciprocal**, the frequency of light**, [1], as **solar constant [W/m²]** or as **extraterrestrial irradiance**, [2] Their value, according to measurement results, is determined by the IAU (International Astronomical Union) on

$$\textbf{1361 [J/m}^2\textbf{s], } [\text{W/m}^2]$$

and is accordingly also managed by CODATA (*Committee on Data for Science and Technology*).

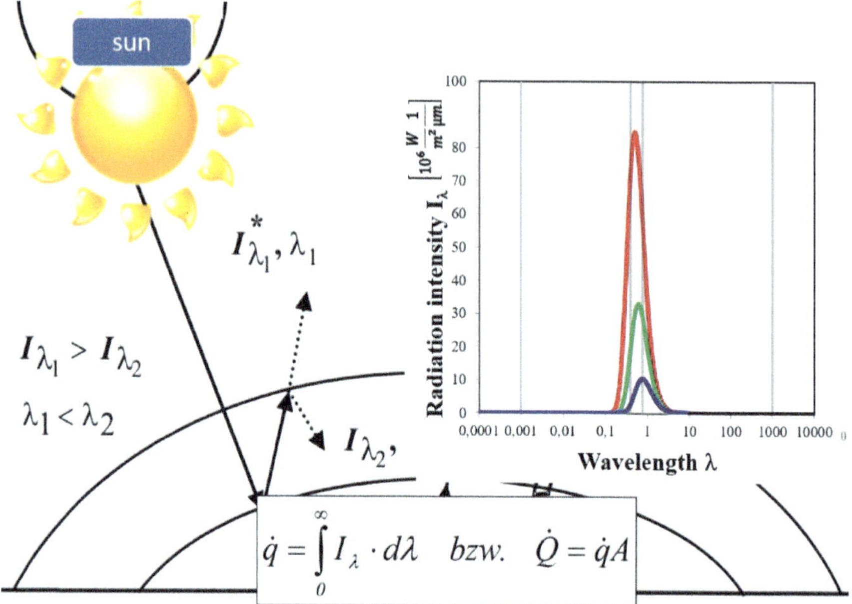

$$\dot{q} = \int_0^\infty I_\lambda \cdot d\lambda \quad bzw. \quad \dot{Q} = \dot{q}A$$

Figure 4.5 Radiation intensity and wavelength of a sunlight radiation for the calculation of the area-related energy flux density [W/m²] and further, if necessary, the energy flux (power) [W]

Basically, this is a thermal radiation that can also be calculated on the basis of the _Stefan-Boltzmann constant_ [1]. Figure 4.6 shows **the energy flux density of solar radiation** (radiation intensity) as a function **of the wavelength** of a beam at different temperatures of the radiator [1].

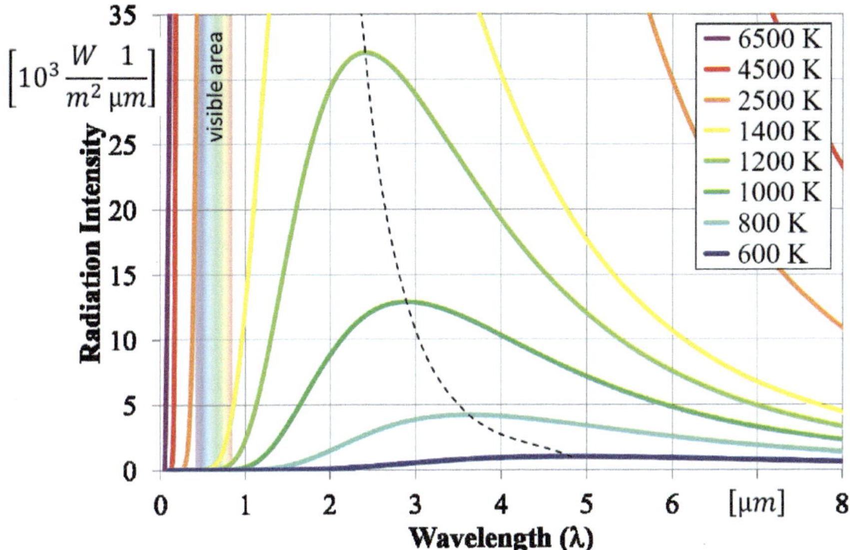

Figure 4.6 Radiation intensity of the sun [W/m³] as a function of the wavelength [1/m] (the sun, as the radiator, *is assumed to be a black body*) at different temperatures of the radiator/according to [1]

Part of the area-related **energy flux** Q [W/m²] penetrates into the atmospheric air, other parts are scattered or reflected [2]. As mentioned at the beginning, the portion of **solar radiation** that enters the atmosphere warms the air, as well as the earth's and ocean surfaces. However, the energy flow coming from solar radiation is also converted directly, in natural processes (photosynthesis in plants) *as well as in technical processes* (photovoltaics in solar systems).

The area-based **energy flux of solar radiation in the Earth's** atmosphere itself changes considerably, depending on the region of the Earth and the time of day or season, fluctuations between 150-1000 [W/m²] [12].

On the other hand, because it is about solar radiation, the distance from the sun to the respective planet also counts, but also the angle of impact of the radiation (ideally vertical). For example, the solar constant [W/m²] is 6.7 times higher on Mars and 1.9 times higher on Venus, but only 0.01 times higher on Saturn [13].

4.3   Solar radiation and the Earth

The flora of this earth itself needs a lot of solar radiation, namely from the part with the highest intensity [W/m³], in the wavelength range [micrometers] of light. Trees and plants use this energy for photosynthesis, which is the engine of their food preparation: with such light rays, chlorophyll, the green hemoglobin in the blood of the plant, is activated,

4.4 Current and Future Materials for Solar Cells which then, in two phases, contributes to the conversion of carbon dioxide from the atmosphere and water at the roots into the nourishing **glucoses**.
Comment:

glucose, *for example in sugar cane or plant residues through fermentation, can also be used to produce alcohols such as ethanol and methanol for combustion engines of vehicles. Brazilians and Americans seem to know a lot about it, they produce millions of flex fuel (ethanol/gasoline variable mixture) vehicles [3]!*

A practical, current example: On an area of 2700 hectares in China **sown with solar panels** (the entire Mosel-Saar-Ruwer wine-growing region extends to 8770 hectares), just 850 *megawatts were* "harvested" in 2018 [14].

850 [MW] divided by 2700 [Ha] results in 3.1 [W/m²] **(is), which appears to be low compared to the planned "should", expressed as the average, area-related power or** energy flux of solar radiation **[W/m²]**. However, the same output is also provided by the oil-fired power plant in Ingolstadt, on 37 hectares, i.e. on 1.37% of the area "sown" with solar panels in China of 2700 [ha].

4.4   Current and Future Materials for Solar Cells Silicon-based Photovoltaic Cell

In 1839, the French physicist Becquerel stated that *"light can cause the release of (electric) charge carriers"*. As a result, an electron can detach from an atom if it has received energetic photons from the sun's ray.

The **upper silicon layer** of a **photovoltaic cell is interspersed** with electron donors, usually phosphorus atoms. When activated by photons from the sunlight beam, with the presence of an abundance of electrons, this results in a detachment in the structure of the upper layer. **The lower silicon layer,** on the other hand**, is interspersed with boron atoms,** where suction effects lead to a lack of electrons, i.e. "holes" in the structure of the lower layer.

However, the process is saturated at some point in this vertical plane, in the upper- and lower-layers electron abundance and electron deficiency in physical boundary of the structure. Then you need more such "shafts" on the horizontal plane, i.e., more area with photovoltaic cells.

This creates an electrical voltage that can be "tapped" through appropriate contacts on the top and bottom. The contacts are connected to a consumer via cables, it can be a *storage battery, a consumer network,* or even *an electric motor.* The electrons from the upper layer then flow through the consumer to the lower layer and so the circle closes again. Then the light rays with energetic photons come through the upper layer to the boundary layer and the game begins again.

Alternative Materials for Photovoltaic Cells

Photovoltaic cells made of **monocrystalline silicon** are by far the most
majority of the systems used in practice. This silicon is produced from quartz sand in electric furnaces at temperatures of more than 1000°Celsius. Photovoltaic cells made of multi crystalline silicon are simpler and cheaper than those made of **monocrystalline silicon,** but they are also less efficient, as mentioned in the example of the solar system of the West Saxon University of Zwickau/Germany. In terms of energy balance, in Ottawa, Canada, a solar system must produce 1.28 years of electrical energy before the energy is amortized during manufacturing and installation. However, in the same view, this is only 0.4 years in India [15]

Other material compositions (**thin-film solar cells**) are listed as follows:

- Cadmium tellurides (CdTe).

- Copper-Zinc-Zinn-Sulfide (Cu2ZnSnS),

- Zinc phosphorite (ZnPh)

In addition, as a more recent development, there is a single-walled carbon nanotube [16], [17] .

Copper Indium-Gallium -Selenide

Copper indium gallium selenide (CIGS) is a thin-film solar cell based on the copper indium selenide (CIS) family.

Perovskite Solar Cells

A perovskite solar cell (PSC) contains a compound, most commonly a hybrid organic-inorganic lead or tin halide-based material which is a light-collecting active layer [18], [19], [20]. **Perovskite materials** such as **methylammonium lead halides and inorganic** cesium lead halides **are inexpensive and easy to produce**.

Dye-sensitized solar cells

Dye-sensitized solar cells (DSCs) are thin-film solar cells that are more effective under ambient light than other photovoltaic technologies. Dye surrounds **TiO2 nanoparticles**, which are located in a sintered network [21].

Many DSCs are made up of liquid electrolytes. The solvent occasionally penetrates most plastics and becomes unstable against temperature fluctuations, causing it to freeze at cold temperatures and expand at warm temperatures. Another problem is that such a solar cell is not ideal for large-scale applications due to its low efficiency.

Some of the advantages of DSC are that it can be used in a wide range of lighting conditions (including cloudy conditions), has low production costs, and does not degrade under sunlight, giving it a longer lifespan than other types of thin-film solar cells.

Organic Photovoltaics

Organic photovoltaics (OPV) is another form of organic, dye-sensitized quantum dot photovoltaics. However, this dye-sensitized photovoltaics poses storage problems: the liquid electrolyte is toxic and can penetrate the plastics used in the cell [22], [24], [24].

The traditional OPV cell structure layers consist of a semi-transparent electrode, an electron blocking layer, a tunnel contact, a hole barrier layer, and, finally, another electrode. where the sun hits the transparent electrode. OPV replaces silver with carbon as the electrode material, which reduces manufacturing costs and makes them more environmentally friendly. They are flexible, lightweight and well suited for mass production. OPV uses *"only existing elements coupled with a low energy through very moderate processing temperatures, using only ambient processing conditions on simple presses that allow energy recovery times"* [25].

Current efficiencies range from 1 to 6.5%. However, theoretical analyses show an efficiency of more than 10 %.

4.5   Temperature influence on the function of solar cells

The performance of a solar panel depends on the environmental conditions. The temperature of the middle part of a solar cell influences the *area-related energy flux density* [W/m²].

This decreases as the temperature increases [26].

This correlation between the output power of a solar cell and the working temperature depends on the semiconductor material and is due to the influence of the ambient temperature on the concentration, lifetime and mobility of the electrons and "gaps".

Temperature sensitivity is usually described by temperature coefficients, each of which expresses the parameter to which it refers in relation to the junction temperature. The values of these parameters can be found in the data sheets of the respective photovoltaic module [27].

Properties Impairment of Solar Modules

Solar panels must be able to withstand damage caused by rain, hail, heavy snow load, heat and cold cycles.

Potential-induced property impairment is a reduction in performance in crystalline photovoltaic modules caused by so-called stray currents [28].

This effect can lead to a loss of performance of up to 30% [29].

This loss of power is mainly due to solar radiation. The impairment index, which is defined as the annual percentage of *output power loss*, is a key factor in the long-term production of a photovoltaic system. To estimate this impairment, the percentage of decrease is associated with each of the electrical parameters. The individual impairment of a photovoltaic module can have a significant impact on the performance of a complete package. In addition, not all modules in the same plant reduce their power at exactly the same rate. In the case of a number of modules exposed to long-term outdoor conditions, it is necessary to consider the individual deterioration of the main electrical parameters and the increase in their dispersion. Each module tends to change its behavior differently, which has a negative impact on the overall performance of the system.

According to a recent study [30], the impairment of crystalline silicon modules varies between 0.8% and 1.0% per year.

On the other hand, when analyzing the performance of thin-film photovoltaic panels, an initial period of severe degradation is observed (which can last several months and up to two years), followed by a later phase in which the degradation stabilizes [31].

With such thin-film technologies, strong fluctuations can be observed over the course of a year. For example, in the case of modules made of amorphous silicon, myomorph silicon or cadmium telluride, annual losses for the first few years are between 3% and 4 % [32].

4.6   Use of photovoltaics in the world

Very large-scale and very small photovoltaic systems are being installed in the world. In order to always put them in the best light, their suppliers and other stakeholders use "bombastic" units of measurement appropriate to the size, which at first acts as a fright for non-initiates: for the power: kilowatts, *megawatts, gigawatts;* for the energy that results per year: *kilowatt-hour, gigawatt-hour, terawatt-hour, but also exajoules and petajoules,* and even *megatons of oil equivalent* (*1 MTOE is the same as 11.63 terawatt hours*).

No one has started with quadrillions of kilocalories yet, but it's never too late! It takes some time and effort to centralize, standardize and evaluate such data according to objective criteria. Even with pure monocrystalline silicon, the efficiency of the described process is currently only (exceptionally) higher than 24%.

So you would need *4.17 square meters of panels* to achieve 1 kilowatt of power *at maximum solar radiation.* In practice, however, it is *5 to 10 square meters,* the silicon crystals are not absolutely monocrystalline, and the cheaper amorphous modules are often used.

Compared to the **calorific value** *of a fuel such as petrol, diesel or ethanol, this is very poor!*

A first, very pragmatic criterion is the ratio of the energy collected in a year, in wind and weather, through summer and winter, through day and night, to the installed and hoped-for (target) output of the system.

In 2018, there were photovoltaic systems in the world with a peak output of 500,000 [**MW**] calculated under standard conditions. However, they produced an actual, total electrical energy of 600 billion [**kWh**] in that year.

In comparison, in 2018, coal produced 9,851 billion [kWh] in the world and gas 5,890 billion [kWh], for a total of 15,471 billion [kWh] with coal and gas. The electrical energy generated by photovoltaics therefore accounts for 3.8% of the electrical energy from coal and gas.

Compared to all energy sources used for electricity production, it is then only 2.3%!

But there is also another interesting aspect: if the 500,000 [MW] could have been reached continuously throughout the year, the total energy produced would be 4380 billion **[MWh]**.

600 billion kWh (actual) divided by 4380 billion **[MWh] (target) results in an efficiency of 0.137!**

Comment:

The solar radiation on all photovoltaic systems in the world, which fluctuates over the day, latitudes and seasons, results in an annual average global radiation efficiency of 13.7%.

By way of comparison, in the same year 2018, there were photovoltaic systems in Germany.
Plants with a peak output of 45,500 **[MW] calculated under standard conditions. In that year, they generated a total electrical energy of (real) 45.7 billion [kWh] in Germany. The radiation efficiency in Germany, in 2018, is therefore 11.4%. This is less than the global efficiency of 13.7%, which can be explained by latitude.**

As a result, the largest photovoltaic systems in the world have been built in very sunny regions.

In **Benban, Egypt,** a photovoltaic plant with a calculated peak output of (target) 1800 [MW] was built in the desert near Aswan [33].

In **Abu Dhabi**, a plant with a calculated peak output (target) of 1177 [MW] is under construction,

In **Longyangxia, China**, in the desert near the city of Hainan, a photovoltaic system was built on 2700 [Ha], as mentioned, which, according to Chinese information, should secure the supply of electricity for 200,000 homes. Converted to the global radiation efficiency in

2018 as a guideline, this results in an electrical energy of around 700 [kWh] per house and year.

*(The guideline value for Germany is **8,000-14,000 [kWh]** for a single-family house with 4 people, with its own solar system, whereby the peak power (target) is 6 [kW]. However, the radiation efficiency in Germany in 2018 is 11.4%, as shown above). Instead of **52560 [kWh]** from 8760 hours with 6 [kW] continuously, an efficiency of 11.4% results in **only 5992 [kWh] – around 6000 [kWh]**! So let's hope that from now on the sun will shine better and longer?*

In **China** , another large plant is currently (02/2020) being built with a calculated peak output of (targeted) 2000 [MW].

In **Kamath, India,** there is a plant with a calculated peak output of (target) 648 [MW],

in **California, USA,** the Topaz Solar Park with a peak power target of 579 [MW], from 1.7 million solar panels [34], [35].

After the desert, however, photovoltaics is also making its way to the sea: In China, a "floating photovoltaic system" was built with a peak power target of 40 [MW] [36].

The 132,400 panels occupy an area of 93 [ha] on the water. Due to the better cooling of the cells, this system has a higher efficiency than the one on the ground.

In Europe, whether **in Italy** or in **Germany**, there has recently been talk of "***Agro photovoltaics***" [37].

There are already plants with vertical panels in the field, with carrots, potatoes and other vegetables growing in between.

It is even said that they grow better than usual due to the light radiation reflected by the panels. The second European variant is the one with panels at a certain height above the pastures, so that the sheep below can eat their grass undisturbed. Such extravagant individual solutions in areas with intensive agriculture and rather moderate solar radiation, such as in Europe, are interesting showcase projects with too little share of the energy production of the respective country.

Results:

As in the examples shown, it seems sensible to build large photovoltaic systems in deserts and in areas with particularly strong and long-lasting solar radiation, such as in **Australia, Kenya, Egypt, China, Dubai, Indonesia, Nevada, Colombia and Guatemala.**

They can efficiently supply the mostly existing grids in neighboring cities with electrical energy in a power mix via short power grids, regardless of fluctuations throughout the day and depending on the weather.

In addition to the large-scale plants, the individual, in-house plants in Europe and elsewhere in the world make sense and are efficient. A system with about 35 [m²] on the roof of a single-family house in Germany should be able to withstand solar radiation (quote: "per kilowatt peak **(1,000 watts) an average of 8 to 10 square meters of space are** required. *A photovoltaic system with an area of **40 square meters** – and thus **4 kilowatt** peak nominal output)* [38] *would be sufficient, wouldn't it?*

So, according to the same, very generous way of thinking, (for 1000 [W] we need "around"10 [m²] roof area), of the planned 35 [m²] 3.5 [kW] should be created. If, however, only 6,000 [kWh] per year come from a "target" of 6 [kW] – <u>corresponding to 60 [m²] of roof area</u>, then now, with only 58.3% of the roof area, it would be just 3500 [kWh]!

There is no need for comments. Or maybe not:

Comments:

- The fluctuation of the electrical energy produced during the day or over the season is not a disadvantage in itself, but a natural fact from which you can make the best. The placement of a large-scale system close to the large power grid of a city was such a compromise, the small system on the house, completed with a battery with a reasonable capacity, was the other.

- *The idea that with the very low yield compared to an energy source such as ethanol from plant residues, photovoltaic participation in the world's electrical energy can be increased from the current 3% to 30 to 50% is very exaggerated. Of course, you could all the deserts of the world with sowing photovoltaic systems, why not, they are empty anyway and the places of people are usually very far.*

- What happens to the fluctuation of energy over day and night, and how do you transport this energy from the desert to Berlin or New York? There is talk of thousand-kilometer lines - in Germany, for example, this has been debated for a very long time, in connection with wind and solar farms, so far without clear results. People are talking about gigantic batteries – especially in the desert, at 50°C, well, you can put a large power-guzzling air conditioner next to it!

4.7 Hydrogen production based on photovoltaic electricity in the form of climate-friendly fuels

8 [kWh], which can be obtained, for example, from about 50 [m²] of photovoltaic panels, is equivalent to the energy of a **single liter** of gasoline or diesel fuel.

Investors and bankers from all over the world are now rushing to Namibia, the United Arab Emirates and Patagonia (Argentina and Chile) in search of plenty of sunshine for photovoltaic installations, (but also wind for wind turbines) to produce "**green hydrogen**" *through electrolysis of water*, a process that requires clean electricity, i.e. primarily by means of photovoltaic systems.

Hydrogen as a fuel for industry and heating

Hydrogen has been a basic ingredient in industry and fertilizer production for over a hundred years. In particular, it can be used in any industrial process, it can be used for heating and mobility. Hydrogen is easily miscible with natural gas, it is

*The **<u>electrolysis of water</u> to produce hydrogen*** appears to be an extremely advantageous method in the context of climate protection: in the case of hydrogen **as a fuel for industry and heating.**

Hydrogen has been a basic ingredient in industry and fertilizer production for over a hundred years. In particular, it can be used in any industrial process, it can be used for heating and mobility. Hydrogen is easily miscible with natural gas, it is often mixed to 20% in the natural gas network, and recently to 30% [39].

In general, industrial hydrogen is produced from *natural gas* (38%) or other Hydrocarbons - fuel oil (24%), gasoline (18%), ethylene (6.6%) - used traditional processes: catalytic reforming, partial oxidation or pyrolysis. Another classic process is the decomposition of water vapor produced during coal gasification (10%). Most of these processes are associated not only with high energy consumption, but also with significant emissions of carbon dioxide.

The electrolysis of water as shown in Figure 4.7 to produce **hydrogen** appears to be an extremely advantageous method in the context of climate protection: in such a reaction, only oxygen is produced in addition to hydrogen.

*The **<u>electrolysis of water</u> to produce hydrogen*** appears to be an extremely advantageous method in the context of climate protection: in such a reaction, only oxygen is produced in addition to hydrogen. often mixed to 20% in the natural gas network, and recently to 30%.

Some "experts" in the fields of economics or politics see the problem in the energy balance: Electrolysis consumes more energy than is produced from the mass of hydrogen produced. Many users of such systems then also think, in this context, the efficiency (meant in the *thermodynamic sense – <u>as a work input,</u> or in the broadest sense "efficiency" **divided by** the energy output)* which makes no sense, as long as the energy utilized, especially that of the sun, is a permanently unlimited source!

Example:

*Four times more energy in electrolytic production or in the production of hydrogen by means of **electrically driven fuel cells** simply means more energy input from the solar cell, which simply needs more solar modules for the same area-related energy of the solar modules. If not intensively, then extensively.*

The fact that systems of this kind, which are scalable to produce smaller or larger quantities of hydrogen, have recently been popping up like mushrooms after rain shows that this is also economically viable.

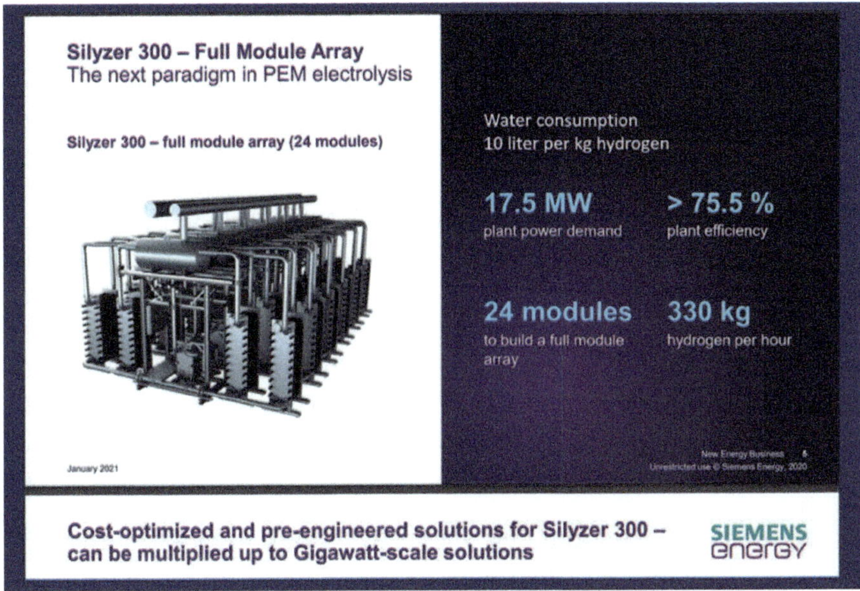

Figure 4.7 Up-to-date, modular and scalable water electrolysis system manufactured in a wide range of Siemens Energy (*Source: Siemens Energy*)

Hydrogen as a fuel for internal combustion engines and fuel cells in mobility

The use of hydrogen as a direct fuel in standard combustion engines or via fuel cells is, according to the examples listed, desirable, but not yet clear for the future. Here are some examples of solutions and trends [40], [41], [3].

In automobiles, hydrogen is already used in fuel cells (*Toyota Mirai and Hyundai Nexo*) and in internal combustion engines (BMW Hydrogen 759 hl series or 760Li).

In 2000, BMW presented a series of 15 such vehicles with hydrogen-based combustion engines.

In 2011, three Mercedes B-Class pre-production fuel cell vehicles (*as a result of the NECAR-New Electric Car hydrogen fuel cell vehicle program initiated by Mercedes in 1994*) covered a distance of 30,000 kilometers on several continents in 125 days (*F-Cell World Drive*), thus proving the feasibility of this technical solution. 200 units of this type were leased to customers selected by Daimler.

BMW is reappearing with such a solution in the iX5 hydrogen fuel cell car, which is to be launched on the market in a limited edition starting this year.

Regardless of the drive solution, fuel cell or piston engine, the price of hydrogen has risen in recent years and is currently just under 13 euros per kilogram in Germany. The price seems very high compared to gasoline or diesel, but it should be noted that a medium-sized fuel cell car consumes about one kilogram of hydrogen per hundred kilometers according to the current cycles. In the case of petrol or diesel, the cost per 100 kilometers is usually higher, at least in Western Europe.

In the areas of heavy-duty engines, heavy-duty trucks or buses, the solutions for the use of hydrogen between fuel cells and piston engines will be divided again, considering costs, complexity, reliability and customer acceptance.

Germany's Deutz AG, a company with sales of 1.6 billion euros, will launch a 7.8-liter 6-cylinder TCG7.8 H2 hydrogen piston engine next year, which can be adapted for a wide range of applications in cars, vehicles and ships.

MAN INC has recently developed a dual-fuel V12 engine that can run on both diesel and hydrogen, designed specifically for large ships. FEV Germany has developed a similar project.

Bus and truck prototypes from the start-up Keyon, Munich, based on the 7.8-litre engine from Deutz AG (left) and the Hyundai hydrogen fuel cell bus.

Hydrogen as the basis for other fuels in industry, heating and mobility.

An alternative variant, the subject of many developments in well-known companies, is the "packaging" of hydrogen so that it can be transported to Europe, the USA, Canada or China.

Hydrogen does not necessarily have to flow through pipelines from where it was produced, like methane gas, to the consumer, who is usually thousands of kilometers away.

This energetic hydrogen can be wrapped very well in pods. Hydrogen, with the lowest molecular weight of all elements, has the lowest density (*15 times lower than that of air!*) at comparable pressure and temperature values. Therefore, for a high density, we increase the pressure or lower the temperature. *Toyota, Hyundai* and other manufacturers of fuel cell cars have produced pressurized tanks – 300, 500, 600 or even 900 *bar (the air in a car's tires, by comparison, has a pressure of only 2.5 bar)*. BMW initially preferred the cryogenic variant (temperatures below minus 253 degrees Celsius), in which hydrogen becomes liquid, but with a density well below that of gasoline or diesel, partially offset by a higher calorific value. However, storage under pressure "costs" about 12% of the initial energy of this hydrogen mass, and cryogenic storage achieves energy consumption of up to 30%.

An alternative solution is to store hydrogen in batteries with **metal structures**, such as sponges that absorb hydrogen into pores, with rather weak chemical bonds that can be dissolved for the subsequent release of hydrogen by increasing the temperature.

However, the most promising solution is hydrogen bonding through tighter chemical bonds, resulting in increased density in the form of **ammonia**. Ammonia is formed using the classic Haber-Bosch process at 300 bar and 450 degrees Celsius. Ammonia liquefies at minus 33 degrees Celsius, as well as at pressures above 9 bar. To "dissolve" such a bond in the components of ammonia, nitrogen and hydrogen, at the destination, it is enough to heat them to 700 degrees Celsius. Liquid ammonia has a density very close to that of gasoline.

Its calorific value is almost half that of gasoline. But while we're at it, what could be a piston engine, a jet engine, or a fuel cell reactor, why not burn it outright?

When ammonia is burned, nitrous oxide, also known as "nitrous oxide", can be produced under certain conditions of the chemical reaction.

So let's stick with pure hydrogen or package it differently:

In LOHCs (Liquid Organic Hydrogen Carriers), which are chemical structures with carbon chains bonded in wire mesh to hydrogen atoms passed through meshes. The hydrogen to be transported can be "bound" with the help of a catalyst using a new technology at 30-50 bar and 150-200 degrees Celsius and then "unlocked" by heating it to over 250 degrees Celsius.

However, one of the solutions with great potential for the future is the use of hydrogen in the formation of **HVO** *(Hydrotreated Vegetable Oils)* **fuels,** which not only hydrogenate plant residues, but also oils that have already been burned, for example from **French fries, or lard and rancid bacon**, which converts them into fuel very similar to diesel fuel, which is already certified in many countries around the world in full or in various blends with diesel fuels for truck engines, as well as for buses, ships and aircraft.

The production of HVO *(hydrogenated vegetable oils)*, which is largely carried out in previous oil refineries due to the strong similarity of the chemical processes, is based on the introduction of hydrogen molecules into the fat or oil molecule. This process practically means a <u>reduction in carbon compounds.</u> Hydrogen, combined with triglycerides by chemical reaction, gives several types of reactions and resulting products [42]. The second step of the process involves the conversion of the triglycerides/fatty acids into hydrocarbons by removing oxygen as water and/or decarboxylation (removing oxygen as carbon dioxide).

A formulaic example of this is

$$C_3H_5(RCOO)3 + 12H_2 \rightarrow C_3H_8 + 3RCH_3 + 6H_2O$$

A supplier who already sells HVO 100 in the greater Ingolstadt area [43] assumes an average surcharge of 15 cents per liter.

Regarding the direct propulsion of cars with **hydrogen as an energy source**, in internal combustion engines and in fuel cells**, there is an extensive and detailed literature** [40], [41], [42], [43].

Comments:

- So the sun comes to the atmospheric boundary with a charge of **1361 [W/m²]**, within the atmosphere only **1000 [W/m²]** remain, the solar panels keep only **100-200 [W/m²]** out of it, over the year, with all fluctuations, only **12-14 [W/m²]** remain of it. From a thousand to twelve, that's 1.2%. However, we need this 1.2% for heating homes and for the mobility of vehicles on roads, through the air and on the seas, in order to relieve the world of anthropogenic carbon dioxide emissions as much as possible.

- In the deserts of the world, as an electrical energy supplier in nearby cities and on family homes all over the world, as a decentralized power supply, photovoltaics is particularly efficient. But covering entire cities and fields with it would neither increase efficiency nor be humane. This can be used to pragmatically produce climate-friendly methanol or HVO decentral, in small or large plants.

- In individual, very modest houses far away from any civilization - there are thousands and thousands of millions of them, whether in Africa, Asia, Australia or even in Europe - a small photovoltaic system on (or as) a roof, connected to a battery, is a real reason to push this direction of development: light, running water thanks to the electrically driven water pump, and the possibility of communication via the Internet in the computer and speaking and seeing via telephone with the rest of the world!

References for Chapter 4

[1]   Stan, C.: Thermodynamics for Mechanical Engineering and Vehicle Construction, 4th edition, Springer, 2020, ISBN 978-3-662-61789-2

[2]   Hans-Günther Wagemann, Heinz Eschrich: *Fundamentals of Photovoltaic Energy Conversion (= Teubner Study Books Physics)*. Teubner, Stuttgart 1994, ISBN 3-519-03218-X

[3]   Stan, C.: Alternative Propulsion for automobiles, Springer, 2020, ISBN 978-3-662-61757-1

[4]   https://www.google.com/search?q=photovoltaik+als+r%C3%BCckgrat+der+elektroenergie&rlz=1C1PRFI_enDE851DE851&oq=photovoltaik+als+r%C3%BCckgrat+der+elektroenergie&aqs=chrome..69i57j0i54614.19660j0j15&sourceid=chrome&ie=UTF-8.

[5]   WHZ Faculty of Electrical Engineering, unpublished material

[6]   https://www.mdr.de/wissen/photovoltaik-solarzellen-wirkungsgrad-neuer-weltrekord-100.html

[7]   Nelson, J.: *The physics of solar cells*. Imperial College Press, 2003, ISBN 1-86094-349-7

[8]   Auto Bild, Series 2022, 2023

[9]   https://www.adac.de/rund-ums-fahrzeug/elektromobilitaet/tests/stromverbrauch-elektroautos-adac-test

[10] SolarWaterWorldAG

[11] solarenergie.de/hintergrundwissen/solarenergie-nutzen/welte-nergiebedarf

[12] Solar radiation in Germany s 4 in Wiki Solar Constant

[13] https://www.astro.uni-jena.de/Teaching/Praktikum/pra2

[14] https://xpert.digital/wussten-sie-es-die-groessten-solarparks-der-welt

[15] Fthenakis, V. M., Kim, H. C. & Alsema, E. (2008). "Emissions from photovoltaic life cycles". Environmental Science & Technology. **42** (6): 2168–2174. *Bibcode*:*2008EnST...42.2168F*. *doi*:*10.1021/es071763q*. *hdl*:*1874/32964*. *PMID* *18409654*. *S2CID* *20850468*

[16] Collier, J., Wu, S. & Apul, D. (2014). "Life cycle environmental impacts from CZTS (copper zinc tin sulfide) and Zn_3P_2 (zinc phosphide) thin film PV (photovoltaic) cells". Energy. **74**: 314–321. *doi*:*10.1016/j.energy.2014.06.076* www.simcoa.com.au. Simcoa Operation

[17] Celik, I., Mason, B. E., Phillips, A. B., Heben, M. J., & Apul, D. S. (2017). Environmental Impacts from Photovoltaic Solar Cells Made with Single Walled Carbon Nanotubes. Environmental Science & Technology

[18] Manser, Joseph S.; Christians, Jeffrey A.; Kamat, Prashant V. (2016). "Intriguing Optoelectronic Properties of Metal Halide Perovskites". Chemical Reviews. **116** (21): 12956–13008. *doi*:*10.1021/acs.chemrev.6b00136*. *PMID* *27327168*

[19] Hamers, Laurel (26 July 2017). "Perovskites power up the solar industry". Science News

[20] Kojima, Akihiro; Teshima, Kenjiro; Shirai, Yasuo; Miyasaka, Tsutomu (6 May 2009). "Organometal Halide Perovskites as Visible-Light Sensitizers for Photovoltaic Cells". Journal of the American Chemical Society. **131** (17): 6050–6051. *doi*:*10.1021/ja809598r*. *PMID 19366264*

[21] Goodson, Flynt (2014). "Supramolecular Multichromophoric Dye Sensitized Solar Cells"

[22] Goodson, Flynt (2014). "Supramolecular Multichromophoric Dye Sensitized Solar Cells"^ Espinosa, Nieves; García-Valverde, Rafael; Urbina, Antonio; Krebs, Frederik C. (2011). "A life cycle analysis of polymer solar cell modules prepared using roll-to-roll methods under ambient conditions". Solar Energy Materials and Solar Cells. **95** (5): 1293–1302. *doi*:*10.1016/j.solmat.2010.08.020*

[23] García-Valverde, R.; Miguel, C.; Martínez-Béjar, R.; Urbina, A. (2009). "Life cycle assessment study of a 4.2k Wp stand-alone photovoltaic system". Solar Energy. **83** (9): 1434–1445. *Bibcode*:*2009SoEn...83.1434G*. *doi*:*10.1016/j.solener.2009.03.012*

[24] Espinosa, Nieves; Lenzmann, Frank O.; Ryley, Stephen; Angmo, Dechan; Hösel, Markus; Søndergaard, Roar R.; Huss, Dennis; Dafinger, Simone; Gritsch, Stefan; Kroon, Jan M.; Jørgensen, Mikkel; Krebs, Frederik C. (2013). *"OPV for mobile applications: An evaluation of roll-to-roll processed indium and silver free polymer solar cells through analysis of life cycle, cost and layer quality using inline optical and functional inspection tools"*. Journal of Materials Chemistry A. **1** (24): 7037. *doi*:*10.1039/C3TA01611K*

[25] Bauer, Thomas (2011). *Thermophotovoltaics*. Green

[26] Piliougine, M.; Oukaja, A.; Sidrach-de-Cardona, M.; Spag-
      nuolo, G. (2021). *Temperature coefficients of degraded
      crystalline silicon photovoltaic modules at outdoor con-
      ditions*. pp. 558–
      570. *doi*:*10.1002/pip.3396*. *S2CID 233976803*

[27] Piliougine, M.; Oukaja, A.; Sidrach-de-Cardona, M.; Spag-
      nuolo, G. (2021). *Temperature coefficients of degraded
      crystalline silicon photovoltaic modules at outdoor con-
      ditions*. pp. 558–
      570. *doi*:*10.1002/pip.3396*. *S2CID 233976803*

[28] *"Solarplaza Potential Induced Degradation: Combatting a
      Phantom Menace"*. www.solarplaza.com

[29] (www.inspire.cz), INSPIRE CZ s.r.o. *"What is PID? —
      eicero"*. eicero.com

[30] Piliougine, M.; Oukaja, A.; Sánchez-Friera, P.; Petrone, G.;
      Sánchez-Pacheco, J.F.; Spagnuolo, G.; Sidrach-de-Car-
      dona, M. (2021). *Analysis of the degradation of single-
      crystalline silicon modules after 21 years of operation*.
      Progress in Photovoltaics. Vol. 29. pp. 907–919.
      doi:*10.1002/pip.3409*. *S2CID 234831264*

[31] Piliougine, M.; Oukaja, A.; Sidrach-de-Cardona, M.; Spag-
      nuolo, G. (2022). *Analysis of the degradation of amor-
      phous silicon-based modules after 11 years of exposure
      by means of IEC60891:2021 procedure 3*. Progress in
      Photovoltaics. Vol. 30. pp. 1176–1187.
      doi:*10.1002/pip.3567*. *hdl*:*10630/24064*. *S2CID 248487
      635*

[32] Piliougine, M.; Sánchez-Friera, P.; Petrone, G.; Sánchez-
      Pacheco, J.F.; Spagnuolo, G.; Sidrach-de-Cardona, M.
      (2022). "New model to study the outdoor degradation
      of thin-film photovoltaic modules". Renewable Energy.
      193: 857–869.
      doi:*10.1016/j.renene.2022.05.063*. *S2CID 248926054*

[33] earthobser-vatory.nasa, 10/2019

[34] https://www.spiegel.de/wissenschaft/natur/aegypten-solarpark-benban-gehoert-zu-den-groessten-der-welt-a-1293132.html

[35] https://www.greentechmedia.com/articles/read/Solar-Star-Largest-PV-Power-Plant-in-the-World-Now-Operational

[36] https://www.ingenieur.de/technik/fachbereiche/energie/welt-groesstes-schwimmendes-solarkraftwerk-in-china-betrieb-aufgenommen/

[37] https://www.ise.fraunhofer.de/de/presse-und-medien/presseinformationen/2019/agrophotovoltaik-hohe-ernteertraege-im-hitzesommer.html

[38] https://www.solaranlagen-portal.com/photovoltaik?ref=566600&gclid=EAIaIQobChMI76nt0cWUgwMVh4ZoCR1iiAbWEAAYASAAEgLiKfD_BwE

[39] https://www.enbw.com/unternehmen/eco-journal/wasserstoff-im-erdgasnetz.html#:~:text=Seit%20November%202021%20wird%20die,Haushalte%20ebenfalls%

[40] Helmut Eichlseder, Manfred Klell, Alexander Trattner: *Hydrogen in Vehicle Technology – Production, Storage, Application*. 4th Edition. Springer, Wiesbaden 2018, ISBN 978-3-658-20446-4

[41] Richard van Basshuysen (Hrsg.): *Gasoline engine with direct injection and direct injection: petrolfuels, natural gas, methane, hydrogen,*
4th edition, Springer, Wiesbaden, 2017.
ISBN 978-3-658-12215-7, p. 520f

[42] Zeman, Petr; Hönig, Vladimír; Kotek, Martin; Táborský, Jan; Obergruber, Michal; Mařík, Jakub; Hartová, Veronika; Pechout, Martin (April 2019). "Hydrotreated Vegetable Oil as a Fuel from Waste Materials". *Catalysts*. **9** (4): 337. doi:10.3390/catal9040337. ISSN 2073-4344

[43] https://www.zieglmeier.de/tanken/hvo_diesel/

Chapter 5

Wind turbines for electricity and mobility

Wind in the sails of brigantines, frigates and corvettes, on all seas, in the windmills of Don Quixote against technical progress in the 17th century.

However, the wind generates energy flows, because of its speed to the power of three. Not far from Australia's coast, the wind blows at up to 406 kilometers per hour, in Patagonia the wind turbines turn 270 days a year at full load. But the wind can also stand still from now on, it can change direction with one hundred and eighty degrees, the wind in Scandinavia, Alaska and on the North Sea coast is always powerful, flat in Mali, Niger and Munich. The "nominal power" has been introduced as a planned "*target*", or like "*W-peak*" in photovoltaics, and it is stated everywhere. The nominal output of all wind turbines in the world sounds enormous at 906 terawatt hours, but this only accounts for 7.6% of global electricity production. In order to mitigate the striking difference between the "targeted" power and the real value, the *full-load* hour was then introduced, then refined as an annual utilization rate or as a capacity factor. The full-load hours are a weak 20% (10-year average in Germany).

The physically correct representation of the work recorded by the wind turbine and then passed on to the electric generator, as a direct dependence on the instantaneous wind speed, goes beyond thermodynamic quantities such as **internal energy** and **enthalpy.** From the balance of these variables, because the wind cannot stand still after a wind turbine, a "*power coefficient*" is derived. And this is only 60%, despite the optimization of the flow profiles of the bucket and its directional flexibility. *And the **power coefficient**, thermodynamically determined, is one thing, the **full-load hours**, whether the wind is stationary or blowing, is another.*

There were nine thousand windmills in Holland in the 19th century, some with outputs of up to **30 kilowatts**. Since then, thirty thousand on-shore (onshore) wind turbines have sprung up like mushrooms from the ground, and five thousand five hundred offshore (offshore).

There have been 1650 full-load hours in recent years, but 2250 full-load hours are expected in the future. In the USA, it is already 2600 to 3500 full load hours. In the current off-shore wind farm Walney, with 102 wind turbines, one million kilowatt hours are generated. *However, a single nuclear power plant can do 166 times as much!* Nevertheless, wind turbines currently generate a fifth of the country's electrical energy.

However, energy production without anthropogenic carbon dioxide emissions absolutely needs the power of the wind, which is present everywhere and again in nature.

5   Wind turbines for electricity and mobility

5.1   Useful work from wind speed

Wind energy is a proven form of conversion of <u>kinetic energy</u> from the gaseous, flowing medium air into work. It is the <u>driving work</u> on seas, in large and small sailing ships, as well as in windmills or for, directly on site, in machines and devices. With their help, grain was ground into flour, groundwater was pumped to the earth's surface, or sawmills were operated. At present, the most important application is the conversion into <u>electrical energy</u>, via powered electric generators.

In order to <u>convert wind energy into kinetic energy</u> and then into <u>useful work,</u> The first condition is wind, a lot of wind. Gigantic wind turbines and immense wind farms depend on the wind as a basic requirement: How strong, how durable and how stable in direction is the wind at selected locations on the planet?

For example, in <u>Barrow Island,</u> not far from the coast of Western Australia, a wind speed of **406 [km/h]** was measured in 1996. Was that just for a few minutes, did the wind change? For this, you don't build an onshore wind turbine for 4 million dollars per [MW] and more (offshore turbines) per [MW] or even 15 million for the largest on-shore wind turbine, with a height of 300 meters [2].

Figure 5.1 shows a wind map (source: NASA), created from frequent measurements within 10 years at a height of 50 meters, i.e. where the hubs of the wind rotors are usually located.

© The Author(s), under exclusive license to
Springer-Verlag GmbH, DE, part of Springer Nature 2024
C. Stan, *Energy Scenarios for the Future*,
https://doi.org/10.1007/978-3-662-69687-3_5

Tierra del Fuego is in the red stripe, with a high and safe wind yield. Right there, in Haru-Oni ("Land of the Wind") next to Punta Arenas, in Patagonia, Chile, in addition to *Siemens Energy and Porsche, the Italian energy group* Enel and *ExxonMobil* as well as the Chilean energy companies *Gasco,* ENAP and *AME,* the main developer and owner of *the project company* HIF (*Highly Innovative Fuels*) are involved in the "Haru Oni" project. At the site of the pilot plant in Chile, a wind turbine runs at **full load for an average** of 270 days a year. In Germany, on the other hand, it is only around 80 days a year with the same investments due to geographical and meteorological conditions. The utilization rate of the wind turbine in Chile is 74 per cent full-load hours, *i.e., around three and a half times higher than in Germany with* 22 per cent full-load hours for on-shore wind turbines [4].

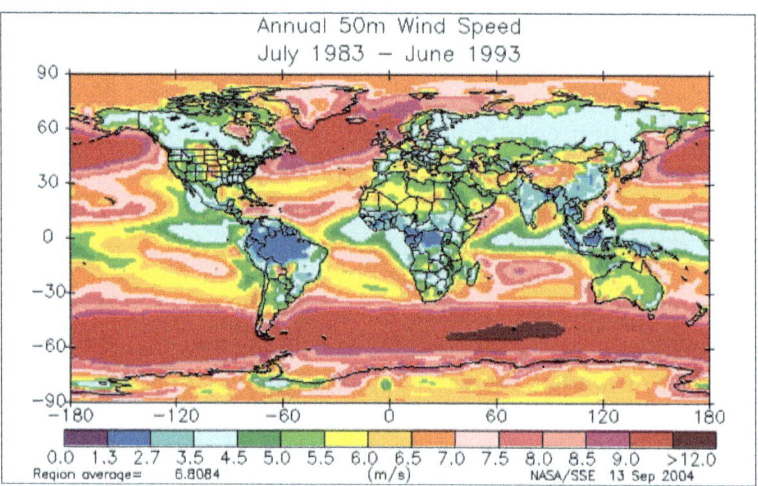

Figure 5.1 Wind map of the Earth, created from frequent measurements within 10 years at an altitude of 50 meters, i.e., where the hubs of the wind rotors are usually located (*Source: NASA*)

The wind off Scandinavia, Greenland and Alaska appears just as intense, and it is more moderate around the equator (Mali, Niger, Chad).

At the end of 2022, wind turbines with a nominal capacity (nominal capacity is a planned "target" size, like W-Peak in photovoltaics) of approx. 906 [GW] were installed worldwide [5].

In 2022, the plants installed worldwide supplied around 2,160 [Twh] of electrical energy, corresponding to about 7.6% of global electricity production [6].

Depending in particular on local wind conditions (often referred to as "*wind efficiency*"), wind turbines reach between 1,400 and over 5,000 full-load hours (the *highest value in the best offshore locations)* [7].

Full **load hours** refer to the time for which a plant would have to be operated at nominal power (*i.e., at optimal target load/full load utilization*) in order to perform the same work as the plant actually did within a defined period of time, during which operating breaks or *partial load operation* may also occur. The **annual utilization rate or capacity factor** derived from the number of full-load hours is the relative full-load usage in a year, i.e. the full-load hours divided by 8760 hours, with 365 days in a year.

The increase in **full-load hours** depends on the technical development of the respective types of power plants, as well as on the geographical and meteorological conditions.

Some examples from 2008: [8]

Energy carrier	Full load hours	Annual utilization rate
Geothermal energy	8300	94,7%
Lignite plant	6650	75,9%
Nuclear energy	7700	87,9%
Wind power Offshore (2011)	2600–4500	29,7 %–51,4 %
Wind power onshore (U.S. new installations 2014)	3600	41,2 %
Water power	4600	52,5 %
On-shore wind power (10-year Central Germany 2016)	1651	18,8 %

In the case of wind turbines, the number of full-load hours is highly dependent on their height, the rotor diameter and the ratio of rotor area to the nominal output of the generator. The average rotor diameter of the new turbines built in the respective year has increased continuously from 22 m in 1992 to over 115 m in 2015 and further growth is possible [9].

Windmills (often referred to as "wind turbines") are approved for certain wind classes in accordance with the international standard IEC 61400 [10].

The IEC wind classes play an essential role in the design of the wind turbine according to windy and low-wind areas. The average wind speeds at hub height are recorded as orientation values. In addition, for the safe construction of a turbine before the rotor blades break, a maximum value of the wind speed (wind extreme value) is considered. This is usually 5 times higher than the statistical average for a region and usually occurs only once every 50 years. The 4 wind classes are:

Class I - 10 [m/s],

Class II - 8.5 [m/s],

Class III - 7.5 [m/s] –

Class IV - 6 [m/s]. (*Reference speed* Vref = Vave × 5)

Figure 5.2 shows the wind zones in Germany as examples [11], [12].

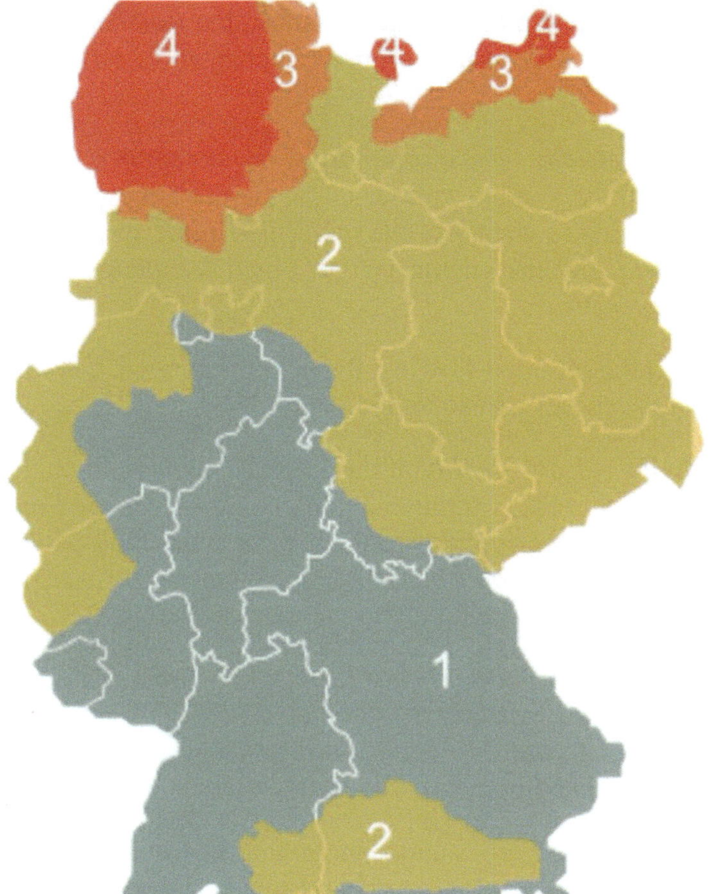

Figure 5.2 Example of wind zones in Germany (WZ1) to (WZ4) (*Source*: [11])

5.2 Wind speed and energy flow for windmill Plants - Fundamentals

The Earth's atmosphere, consisting mostly of air *(about 78% nitrogen, 21% oxygen, then, far below 1% carbon dioxide, which is causing big problems)* envelops our planet, up to about 10 kilometers altitude (*7 km at the poles, 17 km at the tropics*) and is not always as quiet as it seems from far above.

The air has, as determining state variables, the _internal energy_ and the _enthalpy,_ for which the definitions in chapter 3.2 have been presented:

The **internal energy** (_energy quantity of the system in a given state_) - mostly used for systems similar to the Earth's atmosphere - is the sum of all forms of energy of the molecules in the system, first and foremost the molecular kinetic energy, which is expressed externally by the "_temperature_".

However, if the system is flowing, for example as wind, it is advisable to expand the energy consideration:

The **enthalpy** – _is the totality_ of the forms of energy – including the _internal energy_ of the working medium, i.e., the air in the Earth's atmosphere – that have contributed to the system reaching a state. This also includes the "_pumping work_" (p,v) and the _kinetic energy_ of the flow $c^2/2$ itself.

Somewhere in the world, a very large wind turbine is now to be placed in the wind, whether on-shore or off-shore. The turbine is to have a hub height of 135 meters, corresponding to that of the new superlative turbine, Enercon E 126 EP8.

And now, from the Earth's atmosphere, that is, from the air that envelops the planet: From an energetic point of view, before such a wind turbine was considered, there was no further exchange of energy, neither heat nor work – so the enthalpy including the kinetic energy of its velocity (_often referred to as "resting enthalpy, sometimes as enthalpy with dynamic component"_) remained unchanged, at least that's what it looked like from the outside.

In reality, in the flow of air, as shown schematically in Figure 5.3, the proportions within the forms of energy (_specific internal energy u, pumping work (p,v) or specific kinetic energy (c²/2)_) can be expressed throughout the entire dynamic enthalpy h*, which otherwise remains constant, as shown in Eq. (5.1).

$$h^* = u + pv + \frac{c^2}{2} \text{ , or } h^* = \frac{p}{\rho} + \frac{c^2}{2} \text{ , mit } v = \frac{1}{\rho} \qquad \text{Eq. 5.1}$$

Symbol	Unit	Physical quantity
h^*	$\dfrac{J}{kg}$	Dynamic enthalpy (mass-related)
u	$\dfrac{J}{kg}$	Internal energy (mass-related)
p	$\dfrac{N}{m^2}$	Pressure of air in the Earth's atmosphere
v	$\dfrac{m^3}{kg}$	specific volume (mass-related)
ρ	$\dfrac{kg}{m^3}$	Density of air in the Earth's atmosphere
c	$\dfrac{m}{s}$	Velocity of air flow in the Earth's atmosphere

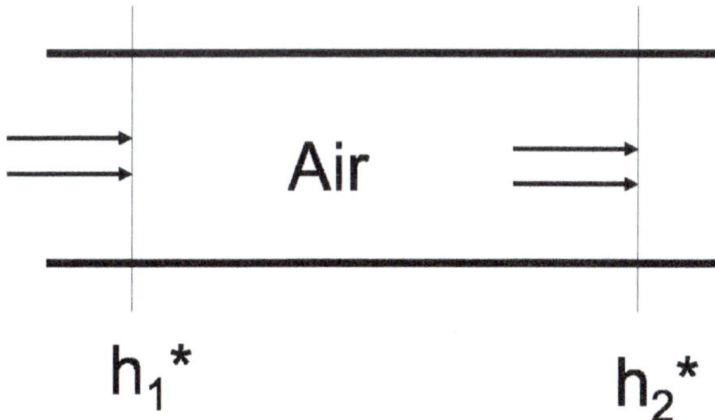

Figure 5.3 Air flow in a virtual, tubular layer

The air density, in connection with the orders of magnitude, is considered to be largely constant. However, the pressure and air velocity are, as usual in practice, variable (Eq.5.2). On the other hand, the specific internal energy of the air, expressed by its temperature, is practically unchanged.

$$h_1{}^* = u_1 + \frac{p_1}{\rho} + \frac{c_1{}^2}{2} \qquad\qquad (\rho_{Air} \cong konst.)$$

Eq. 5.2

$$h_2{}^* = u_2 + \frac{p_2}{\rho} + \frac{c_2{}^2}{2}$$

$$\text{at } h_2{}^* = h_1{}^* \;\rightarrow\; u_1 + \frac{p_1}{\rho} + \frac{c_1{}^2}{2} = u_2 + \frac{p_2}{\rho} + \frac{c_2{}^2}{2}$$

Eq. 5.3

$$\text{with } u_2(T) \approx u_1(T)$$

$$\frac{p_1}{\rho} + \frac{c_1{}^2}{2} = \frac{p_2}{\rho} + \frac{c_2{}^2}{2} \;\rightarrow\; p_1 - p_2 = \frac{\rho}{2}(c_1{}^2 - c_2{}^2) \;\text{ (Bernoulli)}$$

Eq. 5.4

In the equation, as in (Eq. 5.3), the expression for the relationship between velocity and pressure appears in any air flow, known as the "Bernoulli equation" (Eq. 5.4). It is necessary to consider the simplifying conditions of the "Bernoulli" model: no heat from and to the outside (*not looking at the solar radiation at that moment*), no work from and to the outside (no *turbine or compressor in the neighborhood is expected*).

$$c_2 = \sqrt{\frac{2(p_1 - p_2)}{2}} + c_1$$

Eq. 5.5

This causes a change in air velocity, as shown in (Eq. 5.5), which acts in addition to an existing velocity and from a locally existing pressure difference in the atmospheric air.

And now a wind turbine appears in the air current, as shown in Figure 5.4.

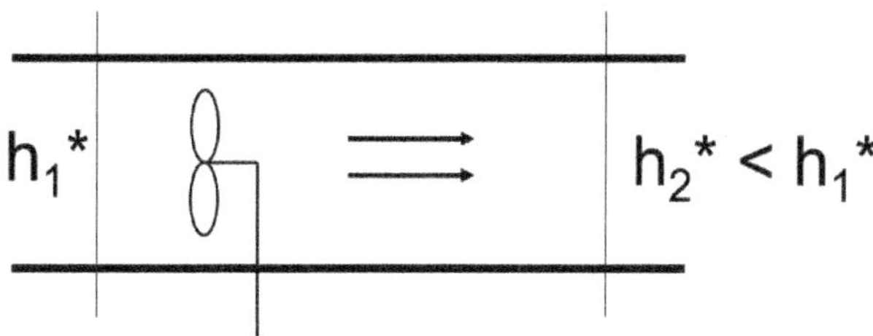

Figure 5.4 Air flow in a virtual, tubular layer in which a wind turbine is located

The wind turbine takes up a work from the air flow, whereby the dynamic enthalpy becomes lower after the wind turbine, according to the First Law of Thermodynamics [1].

The work taken up by a mass flow of air is usually considered to be an energy flow (in this case, a working flow.

Therefore, the consistent representation of the quantities internal energy ($u=U/m$), enthalpy ($h=H/m$), work ($w=W/m$ or heat $q=Q/m$) is advantageous as mass-related. Thus, they remain unchanged for mass flows (Kg/s) instead of masses (kg), except for the designation "energy flow", "heat flow" or "working current". In the case of units, the [J, or kJ] change to [J/s or kJ/s], which quickly leads to the usual designation (J/s=W) and therefore to the term "power.

The "power" of such a wind turbine thus results from the multiplication of the mass-related kinetic energy of the air, caused by an air difference (Eq. 5.5) by the given air mass flow, corresponding (Eq. 5.6).

$$P = \dot{E} = \dot{E}_{kin\ Air} = \dot{m} \cdot \frac{c_2{}^2 - c_1{}^2}{2}$$ Eq. 5.6

with $\dot{m} = \rho A c$

results $\dot{E} = \rho A c \cdot \frac{c^2}{2} = \rho A \cdot \frac{c^3}{2}$

Symbol	Unit	Physical quantity
P	$\frac{J}{s}$, W	Leistung der Luft für die Windkraftanlage
$\dot{E}$	$\frac{J}{s}$, W	Energiestrom der Luft für die Windkraftanlage
A	m^2	Kreisförmige Frontalfläche am Rotoreingang der Windkraftanlage

In Eq.5.6 the influence of the air velocity "to the power of three" can be clearly seen as a decisive factor for such a system. This is the ideal case. But you shouldn't forget the law of conservation of mass: the mass that has come in has to come out, it's like the toothpaste before and after the nozzle, if you press too it, it can be that the pasta splashes into your face.

Eq. 5.6 is given in view of the law of conservation of mass in the form (Eq. 5.7)

$$\dot{E} = \rho A c_i \cdot \frac{c_i{}^2 - c_o{}^2}{2}, \ \cdot c_o > 0 \text{, Law of Conservation of Mass}$$ Eq. 5.7

$$\dot{m}_i = \dot{m}_o \ \rightarrow \ \rho_1 A_1 c_1 = \rho_2 A_2 c_2$$

Symbol	Unit	Physical quantity
c_i	$\frac{m}{s}$	Air velocity at rotor input
c_o	$\frac{m}{s}$	Air velocity at rotor output

Based on the law of conservation of mass, it is not expected that the velocity of the air after the wind turbine will immediately decrease to a standstill. However, the term appears squared in the equation, and is also subtracted. All in all, this makes the yield of the power, or rather the energy flow gained by the air velocity, lower.

This is also known as Betz's Law **since 1919**. The German Albert Betz (1885-1968) put it this way: "*A wind turbine can convert a maximum of 16/27 (almost 60 percent) of the mechanical power that the wind would transport through its projection surface (rotor surface, harvesting surface, effective disc perpendicular to the wind direction) without the braking rotor into useful power*". The factor is also called **the power coefficient** or **degree of harvest**.

According to calculations by various authors, the maximum performance coefficient is approx. 0.5926, i.e., just under 60%.

The above derivation applies to each individual corresponding current tube, so that $v = 2/3 \ v1$ must apply over the entire effective area in order to achieve the global optimum.

For purely historical reasons, the following is mentioned here in this context [13], [14].

„So far, only one theoretically possible way out is known, which has already been stated by Betz himself: If a finite thickness is attributed to the rotor modeled as a single, infinitely thin "active disc", then turbulent fluctuations present transversely to the main flow could introduce additional energy into the volume between the two active discs to be distinguished upstream and downstream. This idea was elaborated in detail by Loth and McCoy in 1983 for a Darrieus rotor with a vertical axis of rotation. They received a power coefficient of 0.62.

The trend is towards higher aspect ratios and longer rotor blades. So that's about 67% to 84% of the theoretical values.

5.3   Use of wind currents in wind turbines

Chapter 5.2 states something that can be expressed numerically:

The instantaneous power of a wind turbine depends on the wind speed to the power of three. Between 5.4 km/h and 25 km/h, the instantaneous power increases a hundredfold with the same air density.

In Germany, for example, four wind zones have been defined according to DIN 1055-4, ranging from a maximum wind speed of 81 km/h (Zone I, Central and Southwestern Germany) to 108 km/h (Zone IV, coasts in northern Germany)!

History of windmills

The first mention of windmills can be found in Babylonian writings from the **18th century BC.** In Europe, windmills were mentioned in England in the 9th century, in France in the 11th century and in the 13th century, and in the 19th century there were already over a hundred thousand of them on the continent, 9,000 only in Holland.

Some of these windmills achieved outputs of up to 30 kilowatts. The first wind turbine to generate electricity appeared in Scotland in 1887. A decade later, aerodynamic wing profiles were developed in Denmark, which greatly reduced the number of rotor blades. The large-scale use of wind turbines began after the historic oil crises of the 1970s.

The reorientation of the energy industry in the direction of renewable energy sources caused by these oil crises in many countries around the world has also significantly boosted the construction of wind turbines, among other measures, such as the very widespread construction of electric cars by large automotive companies.

Since then, onshore wind turbines have been springing up like mushrooms, with around 30,000 of them in Germany alone at the end of 2019 (*Statista 2020*). On the water, i.e., off-shore (off the coast), there are now more than 5500 wind turbines worldwide (11/2019).

On-shore wind turbines are calculated for (targeted) outputs between 2 and 5 megawatts (MW), the off-shore turbines can generally have the better wind in terms of air density and speed, which is why their (target) power values are also higher, i.e., 3.6 to 8 [MW].

However, performance alone is a quantitative criterion, not a qualitative one. The energy efficiency of a wind turbine must be able to compete with that of a photovoltaic cell: If the sun falls at 1000 watts per square meter of cell area, and only 100-200 watts of it can be converted, what about the power provided by the wind on the wind turbine circular surface?

Ratio of power per area

There are widely differing reports on this. This is not surprising, considering that the air velocity is calculated to the power of three, which is why small gusts can have big effects. In a theoretical study from 2015, which is occasionally cited in writings on wind turbines [15], [16], [17], [18], [19], 1 watt per square meter appears as the upper limit.

This takes the wind out of the sails of every wind turbine fan!

Admittedly, this simulation was not made for a wheel, but for an entire wind farm, considering the wind movements over an area of 100,000 square kilometers in Kansas, USA. It can be read that the wind mainly falls from the sky, i.e., vertically, and only swirls on the ground, which is why the horizontal speed components are rather negligible. *If you could hear that in Bremen, Germany, with a crosswind of 108 km/h!!*

And now specifically about the "*full load hours*" defined earlier:

The Walney offshore wind farm, UK, consists of 2 x 51 Siemens wind turbines, each with a capacity of 3.6 megawatts, covering a total area of 73 [km2] on the water. The distance between the individual wheels, which are arranged in rows, is 749 to 958 meters. The total output of all wheels of 367.2 [MW] through the total area of 73 [Km2] would result in 5 [watts] per square meter under continuous full load.

Not only the area in front of each wheel, but also the water area between the wheels was taken into account in order to be able to evaluate the overall efficiency of such a wind farm.

On the other hand, the wind does not blow all year round at the speed that was used as a basis for the maximum power of a wheel. And that's how _the full-load hours came into play: These "full-load_ hours „ during a year, multiplied by the theoretical maximum power, must yield the same energy that was achieved in actual operation with the speed-dependent load fluctuations in all 8760 hours of that year. Depending on the location and turbine design, wind turbines have 1400 to 5000 full-load hours per year. With 8760 hours per year, this results in a utilization rate of 16% to 57% _(the global radiation efficiency in photovoltaic systems was 13.7%)._

On-shore plants in Germany have achieved an average of around _1640 full-load hours_ in recent years. It is expected that in the future the on-shore turbines will have an average of _2250 full-load hours_ and the offshore turbines
4500 full load hours [2]. In the USA, on-shore wind turbines reach _2600 - 3500 full load hours,_ which corresponds to a utilization rate of 30 - 40% [2]. Recently, attempts have been made to compensate for the fluctuations in wind speed by means of _"low-wind turbines"_ with a particularly large rotor area, around 5 m2/kW, which increases the number of full load hours up to about 4000.

Land use: solar systems or wind turbines?

How can we make better use of the land in the future, with solar systems or with wind turbines?

The state-of-the-art _on-shore GE model Cypress wind turbine (02/2020)_ has a maximum calculated (target) output of 5.3 [megawatts] and a rotor diameter of 158 meters, which corresponds to a wind flow area in front of the rotor of 19,596 square meters. This results in 265 watts per square meter. The Vensys wind turbine, with a nominal output of 5.6 [MW] and a rotor diameter of 170 meters, achieves around 240 [_watts per square meter_]. On the other hand, the photovoltaic systems reach 100-200 watts per square meter. The efficiency of solar radiation over the year, comparable as a criterion to the <u>full-load hours</u> of wind turbines, was 13.7% on a global scale in 2018, and 11.4% *in Germany (calculated from the total photovoltaic energy and from the total photovoltaic nominal output in Germany, 2018).*

Compared to photovoltaic systems, wind turbines have both a 1.5 to 2 times higher area-related maximum output and a 1.5 to 2 times higher temporal efficiency (radiation efficiency/full load hours).

If it weren't for the damn nuclear power plants that everyone supposedly doesn't want to have, but still need for their daily electricity!

The Isar 2 nuclear power plant near Munich has a capacity of 1485 megawatts[20] and operates almost continuously (96%) at this capacity, which produced an energy of 12 billion kilowatt hours in 2019. The Walney offshore wind farm has 102 wind turbines, each with 3.6 megawatts, while maintaining an average of 2000 full-load hours per year, resulting in an energy output of 7.2 million kilowatt-hours.

12 billion kilowatt hours with a nuclear power plant compared to 7.2 million kilowatt hours with a wind farm? That's a pretty remarkable ratio from 1667 to 1!

Figure 5.5 Offshore wind farm Walney, UK (Source: Reuters)

The energy generated by the Isar 2 nuclear power plant near Munich per year (decommissioned in 2022) would be offset with 1667 offshore wind turbines, which would be 16 offshore wind farms with 102 wind turbines each, <u>similar to Walney</u> off the coast of Great Britain!

Other comparisons are even sharper: *"If the entire area of Germany were littered with wind farms at a distance of 8 kilometers from each other, the Federal Republic would be able to secure a quarter of its electrical energy,"* according to the daily press. Some self-proclaimed *"specialists love sensation"* similar to *"theorists love simulation"*, sed yes, but reality is often quite different. In Germany, in 2019, more than **21% of the electricity (not the total energy, which is considerably higher, as shown below)** was provided by the already existing 30,000 on-shore wind turbines and the 1464 offshore wind turbines.

Wind energy accounted for over 8% of the world's electrical energy demand in 2019 [21]. However, wind energy accounted for only 0.6% of the world's total primary energy consumption [22].

However, wind farms are increasing worldwide, and energy production without anthropogenic carbon dioxide emissions absolutely needs wind power, along with the other alternatives to fossil fuels.

According to *Statista*, the share of wind turbines in the production of electrical energy worldwide increased from 15.5% to 21.7% between 2015 and 2019.

In China, the world's largest emitter of carbon dioxide, with 28%, the expansion of wind energy is also increasing significantly: The installed wind turbine capacity increased by 40% between 2015 and 2019, from 145 [GW] to 217 [GW], in Germany, by 24% to over 61 [GW] in the same period. In 2020, the USA had an installed wind turbine capacity of 113 [GW], for 2030 224 [GW] is expected, i.e., roughly the same as in China in 2020 [2].

The wind turbines are really springing up like mushrooms, even though they are so big and so heavy. Presented in 2020 by General Electric as the world's largest on-shore turbine, the "Cypress", with a maximum output of 5.5 megawatts, has a rotor diameter of 158 meters, the length of each of the 3 rotor blades is 77 meters. No road train would be able to transport such a rotor blade onto the road to the assembly site, which requires special tractors. The hub height of this wind turbine can be between 101 and 161 meters. Such a colossus needs an appropriate foundation: it weighs 3500 tons, consists of concrete and steel, goes 4 meters into the earth and has a diameter of up to 30 meters. The wind turbine itself also weighs 3500 tons. That makes a total of 7000 tons.

Price

Of course, this also has its price, one speaks of **1600 euros per kilowatt, which would be 8.8 million euros for 5.5 megawatts.** *But the costs don't count, you want to save the world!*

Noise and air vibrations

An interesting aspect is the *rotor speed **of a wind turbine*** in this power range, which, according to various sources, ranges from 4 to 13 revolutions per minute [rpm], depending on the wind speed. At 4 [rpm], the speed of each of the 3 blades at its base, i.e., 2 meters from the center of the hub, is 3 kilometers per hour [km/h]. At the tip of the wing, i.e., 79 meters from the center of the hub, the wing speed is already 120 [km/h]. When the speed is set to 13 [rpm] it becomes 10 [km/h] at the base of the wing but 390 [km/h] at the wing tip, which can make any Formula 1 driver look pale!

The displacement of the air with the three long blades at 200-300 [km/h] provokes not only noise due to lateral friction in the air but also pressure waves due to the frontal contact of the wing edges with the air in front of it: This creates a local air compression, which causes a local increase in pressure. Such an increase in pressure propagates into the ambient air at the speed of sound (at 20°C it is 1235 [km/h]). The whole process develops as a sequence of pressure waves with highs and lows oscillations, with a frequency that depends on the rotor speed. A second source of pressure waves are the vibrations in each blade itself, due to the variable compressive load between the base and the tip, at the speeds, which, depending on the speed and position on the wing, changes from 2 to 120 [km/h] or from 10 to 390 [km/h]. The third source of vibration is the mast itself, which carries the heavy rotating blades at a height of 161 meters. However, these oscillations have a different frequency than those of the air pressure waves in front of the wing edges and then those of the wings over their length. In this way, a wind turbine becomes a small (although so large) symphony orchestra. Some of the people in the vicinity hear such vibrations, and some of them feel them.

Figure 5.6 General Electric's on-shore wind turbine "Cypress", currently (2020) the largest in the world (*Source: General Electric*)

For humans, but also for the animal world, music has <u>a natural key-note</u> defined by its frequency – **432 hertz**, or 25,920 sound wave peaks per minute. This is the resonant frequency of the cells in the bodies of humans and animals. The clear influence on the plants and trees is not discussed here.

Medical studies have identified various influences of the low frequencies caused by wind turbines on humans: On the one hand, they cause clear reductions in the strength of the heart muscle (*Gutenberg University Mainz*), and on the other hand, the activation of various regions in the brain that are responsible for stress (*University Medical Center Hamburg Eppendorf*).

The fact that birds and insects are also exposed to the rotor blades of the wind turbines is, of cause, of interest:
The German Aerospace Center recently determined that 8,500 buzzards, 250,000 bats and 1,200 tons of insects were killed by wind turbines in one year. Opponents, however, believe that the birds eat many more insects, about 400,000 tons in a year in Germany.

There will always be pros and cons in every respect, but especially in connection with promising innovations.

The church should still be left in the village, and please **no giant windmill next to the church**, that not only makes the priest nervous.

References for Chapter 5

[1] Stan, C.: Thermodynamics for Mechanical Engineering and Vehicle Construction, 4th edition, Springer, 2020, ISBN 978-3-662-61789-2

[2] Stan, C.: Energy versus carbon dioxide, Springer, 2021, ISBN 978-3-662-62705-1

[3] https://magazin.schindler.de/technologie/das-groesste-windrad-der-welt

[4] https://media.porsche.com/mediakit/gp-ice-race/de/efuels

[5] https://gwec.net/globalwindreport 2023/

[6] Global Electricity Review 2023 https://ember-climate.org/insights/research/global-electricity-review-2023/

[7] Martin Kaltschmitt, Wolfgang Streicher, Andreas Wiese (eds.): *Renewable Energies. System engineering, economic efficiency, environmental aspects*. Berlin /Heidelberg 2013, p. 819

[8] https://de.wikipedia.org/wiki/Volllaststunde

[9] Berthold Hahn, Volker Berkhout, Bernd Ponick, Cornelia Stübig, Sarina Keller, Martin Felder, Henning Jachmann 2015: The Limits of the growth has not yet been achieved. (PDF) Wind Industry in Germany; accessed 1 August 2016

[10] *Directive for wind turbines. Impact and stabilityverifications for tower and foundation.* (PDF) German Institute for Building Technology

[11] Eurocode 1 (EN 1991-1-4)

[12] DIN 1055-4

[13] Robert Gasch, Jochen Twele: *Wind turbines. Basics, Design, Planning and operation.* 8th edition, Springer, Wiesbaden 2013, ISBN 978-3-8351-0136-4, p. 194f

[14] Alois P. Schaffarczyk: *Introduction to Wind Turbine Aerodynamics.* Springer, Berlin, 2014, ISBN 978-3-642-36408-2

[15] Miller, L.M. et al: Two methods estimating limits to large scale wind power generation, PNAS (Proc. of National Academy of Sciences, USA), September 8, 2015 112 (36) 11169-11174, Edited by Chr. Garrett, University of Victoria, Kanada

[16] Hahn, B. et al.: The limits of growth have not yet been reached, Wind industry in Germany, November 2015

[17] Mills, A.; Wisera, R.; Kevin Porter, K.: The cost of transmission for wind energy in the United States: A review of transmission planning studies. In: Renewable and Sustainable Energy Reviews 16, Ausgabe 1, 2012

[18] Hirschberger, St. et al: Comparative Assessment of Severe Accidents in the Chinese Energy Sector. Scherer Institute, March 2003, ISSN 1019-0643

[19] ***   Annual Energy Outlook 2019, US Energy Information Administration, USA

[20] https://de.wikipedia.org/wiki/Kernkraftwerk_Isar

[21] Angabe von Global Wind Energy Council – GWEC

[22] Angabe von Statistical Review of World Energy, 2017

Chapter 6
Hydropower and hydroelectric power plants

Every flowing water forms an energy flow through its <u>mass flow</u> and through its <u>potential energy</u>, usually represented by the height of the waterfall - *which is converted into* , mechanical energy <u>and further into</u> electricity. A stream of water, including height, leads to hydropower. Basically, the water wheels in the hydroelectric power plants are like the wind turbines. However, water has a thousandfold density compared to air. The height of the fall, in the case of water, sometimes leads to the same speed as that of the air over Patagonia. However, the enormous density significantly changes the orders of magnitude between hydropower and wind power plants: The largest wind turbine, GE Cypress in the Netherlands, has a maximum output of **5.3 megawatts** – the "Three Gorges Hydroelectric Power Plant" in China **22,500 megawatts**.

Strictly speaking, this is not (water) power, but energy per unit of time, comparable to work, and from this also the electrical power. But you build plants, power plants, like coal-fired power plants: in other words, hydroelectric power plants.

The power in electrical, wind and water flows has the same roots: an **intensity** (electric current, mass flow of air *and water) and a* potential (electric voltage, wind speed, water speed due to the difference in altitude). More mass flow or more height difference in water flow is like electricity times voltage in the electrical system: **power**. The diversity of hydropower plants in the world is not limited only to their size, but in particular to the type of combination between large/low mass flow and the flow velocity caused by large or small heads.

For example, **run-of-river power plants** generally work in flowing waters, at relatively low heads, such as at Niagara. **Storage power plants**, on the other hand, combine pressure and speed potential, which is why the turbine type is different at heights between 25 and 400 meters. Naturno, in South Tyrol, reaches a drop height of 1150 meters. Pumped storage power plants manage the heights cleverly, i.e. by giving and realising potential. Recently, there have been micro-hydropower plants on rivers with low hydropower potential, with a water drop height of about 2.5 meters and a speed of five kilometers per hour.

Norway, for example, has 1500 hydroelectric power plants, which provide 93.5% of the country's electrical energy. This partly explains Norway's drive for electric mobility.

Overall, water provides 17% of the world's electrical energy, which means that 70% of the world's clean electricity came from water (2015). In the meantime, however, the use of wind power has increased to almost 21%, while that of hydropower has shrunk from 70% to 58%. There are many reasons for this: huge investment costs, many ecological concerns and, last but not least, the disasters that have not stopped since ancient times. *The dam wall of the largest dam and drinking water reservoir of the Romans, built in the time of Nero (54-68 AD), not far from Rome, collapsed in 1305.*

The Banqiao Dam in China, built in the late 1950s, had a

Capacity of almost 50 million cubic meters of water. The collapse of the Banqiao Dam caused the cascading rupture of another 62 dams. More than a million people were trapped by the water for days, out of reach for any disaster relief. At least 26,000 people died from the immediate tidal waves, and another 145,000 died from starvation and epidemics.

Climate-friendliness sometimes came at a catastrophic price.

6 Hydropower and hydroelectric power plants

6.1 Electrical power from water flow

The energy flow of a **water stream** [J/s, W] converted into **mechanical** *energy (also called power after the units [W])* and then transferred to *an* electric generator **via the rotor of a fixed machine (turbine with blades or pump with rotary- or translating-piston in a hydroelectric power plant)** is a proven form of conversion *of kinetic energy* of a flowing water into *electrical energy.*

So, it seems to make more sense to use a **momentary energy supply instead of hydropower. Work** [W] or, more precisely, to talk *about a* working current [**W= J/s joules per second**], which is nothing more than an *instantaneous power* [**W-watt, kW or Kilowatt**]. Over a certain period of time, for example over a year, you can add up the power [*W-watt*] to each hour, resulting in the **total energy** [*kWh-kilowatt-hour*].

The instantaneous power in the turbine of a hydroelectric power plant is determined in a similar way to that in the rotor of a wind turbine: The basic elements are

- **the mass flow** or**,** if the density is known (general, invariable, with the value, for water 1000 kg/m³),

- **the volume flow** of the water and its energy**:**

In the case of the wind in the wind turbine, this was the kinetic energy expressed by the wind speed,
In the case of water in the **hydroelectric power plant,** this is the potential energy represented by the water head and gravitational acceleration *(for our latitude, with the value 9.81 m/s²),* as shown in Eq. (6.1) to (6.4) [1].

© The Author(s), under exclusive license to
Springer-Verlag GmbH, DE, part of Springer Nature 2024
C. Stan, *Energy Scenarios for the Future,*
https://doi.org/10.1007/978-3-662-69687-3_6

$$\dot{m} = \rho A c \qquad \text{Eq. 6.1}$$

$$H_{dyn} = m\left(u + pv + \frac{c^2}{2} + g\Delta z\right) \qquad H_{dyn} = H^* \, from \, Eq.\,5.2 \qquad \text{Eq. 6.2}$$

$$H_{dyn} = \rho A c\left(u + \frac{p}{\rho} + \frac{c^2}{2} + g\Delta z\right) = Ac\left(\rho u + p + \frac{\rho c^2}{2} + \rho g\Delta z\right)$$

whereby $\rightarrow \rho g\Delta z$ $-$ hydrostatic pressure p_z

$p_{tot} = p + p_z$ $-$ Total Pressure p_{tot} $\qquad$ Eq. 6.3

$$H^*_{\,dyn} = \eta Ac\left(\rho u + p + \frac{\rho c^2}{2} + p_{tot}\right) \qquad \text{Eq. 6.4}$$

Symbol	Unit	Physical quantity
h^*	$\dfrac{J}{kg}$	Dynamic enthalpy (mass-related)
u	$\dfrac{J}{kg}$	internal energy of water (mass-related)
p	$\dfrac{N}{m^2}$	Pressure of water
v	$\dfrac{m^3}{kg}$	specific volume (mass-related)
ρ	$\dfrac{kg}{m^3}$	Density of water
c	$\dfrac{m}{s}$	Speed of water flow
g	$\dfrac{m}{s^2}$	Earth acceleration, latitude-dependent
A	m^2	Flow cross-section of the water
η	$-$	Efficiency of the flow / the system
p_{tot}	$\dfrac{N}{m^2}$	Total pressure of water with hydrostatic fraction p_z

However, there is a huge difference between the kinetic energy of air movement (wind) and the potential energy of falling from a height (water): water has almost a thousand times the density of air, which is why the current power of a wind turbine and that of a hydroelectric power plant appear to be of different orders of magnitude:

The GE Cypress wind turbine has a maximum capacity of 5.3 megawatts. The Three Gorges hydropower plant in China has a maximum capacity of 22,500 megawatts!

Just as long as people in their history have used **fire** and **water to generate** energy **in the form of heat or work, another element, water, has also been used for this purpose**.

The first watermills were used 5,000 years ago in China and Mesopotamia [3], according to the chronicles. Later, in the VI century BC, a hydroelectric power plant was used in Jordan to cut stones during the construction of the Temple of Artemis. The first watermill to produce electricity for domestic lighting was built in England in 1878 [4], [5].

And a short time later, in 1895, the world's first large-scale hydroelectric power plant for the production of alternating current was put into operation above Niagara Falls.

After a few years, the installed capacity reached more than 78 [MW]. In 1961, this hydroelectric power station was replaced by a new one, which now has a capacity of 2,400 [MW] [6].

As long as the considered mass flow of <u>water</u> or <u>air</u> does not exchange additional energy in the form <u>of heat</u> and <u>work</u> with the environment, its total energy (*more precisely:* **dynamic enthalpy***)* remains constant: the mentioned components – **internal energy,** *pumping,* **kinetic energy of the flow** *upon arrival at the turbine,* **potential energy, expressed by the head of the fall,** can exchange energy with each other during a process [1], without changing the overall dynamic enthalpy of the system (*water in motion*).

For the qualitative evaluation of processes of all kinds, whether in wind turbines or in water turbines, the energy, including its main components, is generally related to **the mass** (kilograms) of the respective working medium, water or air. If you compare *the mass flow multiplied* by *the mass-related (specific) energy, the result is an* energy flow *that is nothing other than* the instantaneous power *in the respective component of the working medium (*internal energy).

From this point on, the transformation process begins for the water, as a working medium, on the way to the turbine: the water in such a system does not exchange any further energy with the environment, neither <u>as heat</u> nor as <u>work</u>. The water falls from the mountain because it has a potential energy in the given height of the *fall*. This potential energy is largely converted into *velocity (i.e., kinetic energy)*, a *residual pressure (potential energy)* remains, depending on the conditions on the way. <u>*The internal energy*</u>, expressed by the *temperature* of the water, generally remains unaffected, except for some friction, which is converted into frictional heat, which is usually given to the environment as a loss.

Now the water from the mountain could fall directly into the turbine. However, it would be more advisable to collect it first in a storage tank, as is common in hydraulic systems, in order to avoid pressure fluctuations. In this case, <u>*the hydrostatic pressure* p_z</u> is converted into the <u>*total pressure* p_{tot}</u>. There are several types of turbines. At this point, only one (*the Francis turbine*), Figure 6.1, will be considered as an example, which is most often used in hydroelectric power plants. This radial turbine works like a "*slingshot*" in which the <u>potential energy</u> (*determined by the pressure*) is converted <u>into a kinetic energy</u> *(determined by the velocity* [1].

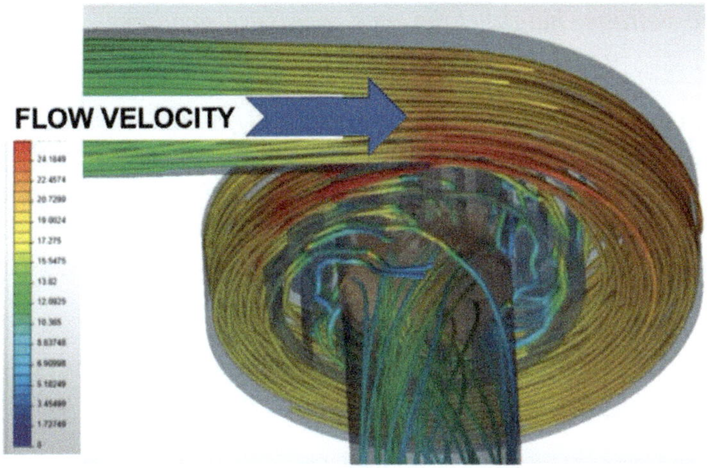

Figure 6.1 Flow path of the water, from the outside to the inside, in a radial turbine
(Francis Turbine)

- In the case of the slingshot, the stone is given a speed by rotation, as external work. With the Francis turbine, the water enters the vortex from the outside at a certain speed.

- By flowing from the edge to the center and by redirecting the direction of the water from horizontal to vertical, towards the exit of the turbine, most of the speed (kinetic energy) is taken away from the water.

- This is converted into pressure (potential energy), which presses on the flanks of the rotor blades. Now, these blades are not fixed, but movable on one axis.

- And so, this potential energy reaches the outside, as torque at a corresponding rotational speed.

- Torque multiplied by rotational speed gives work. And this is given to the power generator flanged to the axle.

<u>Comments:</u>

Power has *the same roots* for **electric, wind** and **water flow**: an *<u>intensity</u> (electric current, mass flow* of air or water) and a *<u>potential</u>*

(*electric voltage, height difference, pressure gradient, speed difference*).

Just as in the *electrical system – power is current times voltage* – wind and hydroelectric power plants are also built: more *mass flow* (according to the intensity in the electrical system) - or more *potential* (pressure difference for wind speed in wind turbines or head of fall in hydropower plants - both correspond to a voltage in the electrical system).

A concrete example shows both the correlations between mass flow and height difference, but also the magnitude of the achievable power:

Ten buckets of water per second (*that's 100 liters, and with the water density of one kilogram per liter, just 100 kilograms*) flow from a height of 10 meters through a pipe into a turbine. Mass flow multiplied by the gravitational acceleration (9.81 m/s²) results in no more and no less than 10 kilowatts of power at the input in the turbine! The 10 kilowatts are reduced by the efficiencies of the turbine, the gearbox, the generator and the transformer, but only to 8.5 kilowatts.

In comparison, the largest wind turbine in the world, "Cypress", produces **0.265 kilowatts** per square meter at maximum power. For the **8.5 kilowatts**, as with the 100 liters of water that fall into the turbine per second from a height of 10 meters, part of the rotor area of **32 m2** is required. On the other hand, a comparison with the best solar cells, with **200 [W/m2]**, an area of **850 [m²]**, almost like the base area of the Reichstag Dome.

6.2   Types of hydropower plants

The diversity of hydropower plants in the world is determined not only by their size, but in particular, by the way in which they combine the greater or lesser mass flow of **the** water as a working medium and the amount of the mass-related (specific) enthalpy (which is usually determined by flow velocity/*kinetic energy or by hydrostatic pressure or pumping/potential energy*) generated [7], [8].

Run-of-river power plants are generally built in flowing waters, with a barrage at a weir. The potential due to a **drop height** is usually low, up to 15 meters, so the power is determined by the **mass flow**, as in the hydroelectric power plant above the Niagara Falls.

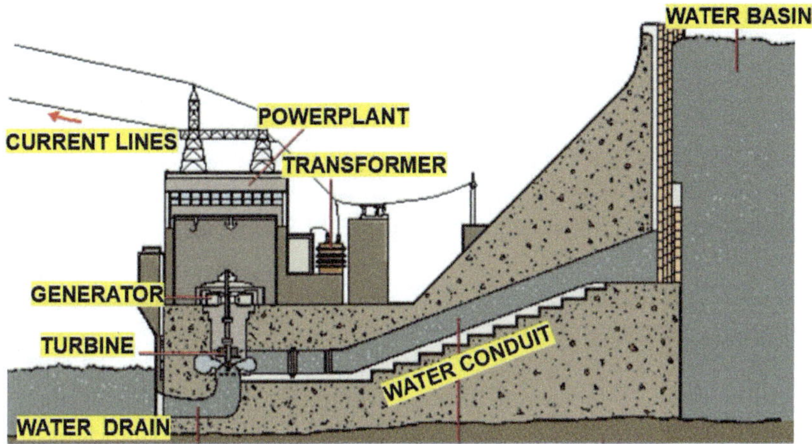

Figure 6.2 Storage power plant (source: [2])

Storage power plants get their water from reservoirs (reservoirs or ponds). Between the reservoir, at a certain depth that provides a pressure potential, and the turbine, a diagonally running water pipe with a defined head is built Figure 6.2. This is, in terms of dynamic enthalpy, as expressed in Eq. (6.2), a combination between the pressure potential at the inflow point and the velocity potential in the conduit. Hydropower plants with heads of 25 to 400 meters cover base and medium loads of the required electrical energy and use Francis turbines in particular, as described above. At higher heads, the pressure potential increases accordingly, as does the pressure on the turbine blades. Therefore, in these cases, not only Francis turbines *(which are inverted "slingshots"),* but also Pelton turbines are used – *(these are "spoons" on a wreath into which the water falls from above).* These are often used to cover peak loads.

The power plant with the largest head in the world is located in Naturno, South Tyrol – it is an impressive 1,150 meters! The plant has a capacity of 180 [MW].

Pumped storage plants are clever in their use of electrical energy: they consist of two water basins at different heights, between which a downpipe system is built. When the economy of the near or further vicinity needs energy, the water flows from the full upper basin via the turbines into the rather empty lower basin, thus driving the power generators and covering the peak loads. And so, at some point, the water level in the upper basin is low and in the lower basin is high. But at night there is no need for electricity to the outside, people like factories have their night's rest. Throughout the night, until the morning hours, the water is pumped up from the lower to the upper basin with the help of electricity from the mains, which almost no one needs at that time. And so, it goes day and night.

The largest hydroelectric power plant in the world, "Three Gorges", with a reservoir length of 663 kilometers, was built in China (completed in 2008) and has a nominal capacity of 22,500 [MW]. The second largest plant, also in China, planned for 16,000 [MW], will be completed in 2021 [9], [10]. Incidentally, a quarter of the world's hydropower capacity is installed in China. Brazil and Paraguay have jointly built a hydroelectric power plant with a capacity of 14,000 [MW] on the Rio Paraná, currently the third largest in the world.

The largest hydropower nation in Europe is **Norway** with 1500 hydropower plants (2019) that secure 93.5% of the country's electrical energy [11].

Water thus provides about 17% of the world's electrical energy! [11]

Hydropower plants, on the other hand, have held the top spot in electricity generation from renewable energies for many years. In 2015, nearly 70% of the world's clean electricity came from hydroelectric power plants, followed by wind turbines at 15.5%. Photovoltaics managed just 5%. However, the economic, technical and ecological developments in the world are leading to tendencies that sometimes seem unexpected:

In the period 2015-2019, the share of wind power in world electricity has steadily increased, reaching almost 21%. Photovoltaics has moved in the same direction during this period, doubling to more than 10% in 2019. The participation of <u>hydropower</u>, on the other hand, shows a thought-provoking decline from 70% to about 58%. The reasons are clear:

The investment costs are very high and would only make such a plant profitable if these costs were covered by electricity sales. But if the state is behind it for ecological reasons, like China, the situation is quite different.

However, there is also a need for discussion and analysis about ecological reasons: Hydropower plants are **atmosphere-friendly** because the energy they produce does not cause carbon dioxide emissions, but that does not necessarily mean they are **nature-friendly**. Reservoirs are a huge intervention in the groundwater balance. As a result, the rivers become unbalanced, and flora and fauna are affected. Entire places of people are often relocated for the construction of dams.

Hydroelectric power plants have always been a <u>disaster hazard</u>:

In China, near Nanking, the largest dam of the time (the dam was 48 meters high and 4500 meters long, the reservoir had 6700 km²) was destroyed by floods in the year 516, there were 10,000 deaths. Admittedly, this dam was not intended to generate electricity, but to flood enemy armies. [12]

The dam wall of the largest dam and drinking water reservoir of the Romans, built in the time of Nero (54-68 AD), not far from Rome, collapsed in 1305.

In France, near Frejus, a dam was built in 1954, creating a reservoir for water supply and irrigation. It broke in 1959 as a result of a tidal wave. 432 people were killed.

In California, 3 dams broke as a result of an earthquake in 1971.

The Banqiao Dam in China, built in the late 1950s, had a capacity of nearly 50 million cubic meters of water. As a result of a typhoon, there was a huge flood that could not be stopped by the lock gates. The collapse of the Banqiao Dam caused the cascading rupture of another 62 dams. The water waves were several meters high and flowed into the lowlands at almost 50 km/h. Numerous villages were flooded so quickly that there was no time left for the evacuation of people and animals. More than a million people were trapped by the water for days, out of reach for any disaster relief.

At least 26,000 people died from the immediate tidal waves, and another 145,000 died from starvation and epidemics [6].

<u>Micro-hydropower plants</u> are more ecologically compatible, less hazardous and more cost-effective than a single, large hydropower plant with a large dam, a large reservoir and a large head [13].

Recently, micro-hydropower plants have been developed for river sites with low hydropower potential. The small mobile power plants operate without accumulating water. All that is required is a water drop height of at least 2.5 meters, with a water width of around five meters and a flow speed of over 5 km/h. For this purpose, new horizontal water wheel variants are being developed. The micro hydropower plants can also be set up as a fleet, which opens up promising paths!

References for Chapter 6

[1]   Stan, C.: Thermodynamics for Mechanical Engineering and Vehicle Construction, 4th edition, Springer, 2020, ISBN 978-3-662-61789-2

[2]   https://www.suewag.com/corp/gruen-und-nachhaltig/strom-aus-wind

[3]   wilo.com/de/de/News-Blog/Rund-um-die-Wilo-Welt/Blog/5000-Years- Hydropower-as-from-the-bucket-wheel-a-hydroelectric power plant

[4]   Jeremy Black: The Making of Modern Britain. The Age of Empire to the New Millennium. The History Press, Chalford 2008, ISBN 978-0-7509-4755-8

[5]   Theodor Strobl, Franz Zunic: Wasserbau. Current fundamentals – new developments. Springer, Berlin/Heidelberg/New York 2006

[6]   https://www.ren21.net/wp-content/uploads/2019/05/GSR2017_Full-Report_English.pdf

[7]   https://web.archive.org/web/20140627215129/http://www.ie-ahydro.org/What_is_hydropower's_history.html.

[8]   Jürgen Giesecke, Emil Mosonyi, Stephan Heimerl: Hydropower plants: Planning, construction and operation. Springer Science & Business Media, 2009, ISBN 978-3-540-88988-5

[9]   Achim Gutowski: The Three Gorges Dam in the People's Republic of China: Background, Cost-Benefit Analysis and Feasibility Study of a Large Project Taking into Account Development Cooperation, Bremen 2000

[10] Lorenz King, Marco Gemmer, Martin Metzler: The Three
      Gorges Project on the Yangtze – Giessen Research
      Group investigates the effects of the world's largest
      dam project. – Mirror of research

[11] https://www.statkraft.de/stromerzeugung/wasserkraft/

[12] Andrew Charles, Paul Tedd, Alan Warren: Lessons from his-
      torical dam incidents. Hrsg.: Environment Agency.
      2011,
      ISBN 978-1-84911-232-1 (englisch)

[13] www.elektroniknet.de/power/guenstiger-als-wind-und-
      sonne.206877.html

Chapter 7

Nuclear energy for heat and electrical energy: climate-friendly, sufficient, but questionable

Around 1950, nuclear energy already accounted for 15-18% of the electricity supply, but this decreased to 10% for reasons of various kinds, from *safety, disposal of nuclear waste, but above all because of the enormous costs during construction*. Nevertheless, their share remains above that of *wind and solar energy* or *photovoltaics*. Energy production without any emission of carbon dioxide and stable energy supply in the grid, without fluctuations, as with wind or photovoltaic energy generation. Countries with a solid economy have a correspondingly enormous energy demand, which can hardly be met without nuclear energy. Of the 250 nuclear power plants worldwide, the USA operates 92 reactors in 54 nuclear power plants (2022), France 56 reactors in 18 nuclear power plants, Japan (!) 33 nuclear reactors (August 2023) with 27 decommissioned reactors. Almost all nuclear power reactors in the world are pressurized *water reactors (in the past, boiling water reactors were also built*, with a circuit), with two separate circuits: in the first circuit, the water in the "bath oven" is heated to about 300 degrees Celsius with the help of the nuclear reaction. Admittedly, with coal or wood, like grandma's, the temperature would be almost twice as high, but this would emit carbon dioxide and other gases. A nuclear reaction emits nothing at all, except rays, which tend to become quite a problem later on.

The modules and the process sequence in the second circuit can then be seen in exactly the same way as in a coal-fired power plant: with hot steam in the boiler, a steam turbine can then be provided with work that rotates a power generator. Due to the maximum temperature to which the nuclear reactor is limited for safety reasons, the thermal efficiency remains at 33 to 40%, i.e. below that of a modern coal-fired power plant.

It is assumed that a nuclear power plant will have a service life of 60 years. On average for all reactors in the world, it is practically half. So, it is not advisable to build 50 reactors almost the same in one country. The reconstruction is then almost the same, which can cause power grid problems. The cost of a nuclear power plant and, implicitly, the price of electricity per kilowatt hour is a matter for the observer:

At four billion euros (*up to 1600 euros/megawatt hour in one example, in Lithuania),* nuclear electricity, including construction investment, waste disposal and shutdown, is the most expensive form of electrical energy production, according to the Rocky Mountain Institute (2005). Without construction and decommissioning costs, simply during operation, this consideration results in the cheapest electricity, says British Petroleum (also 2005).

The key problem, however, is reactor safety and nuclear waste disposal. Germany, Belgium, Switzerland and Taiwan want an early shutdown or have already achieved it. Italy and Egypt no longer want to invest in nuclear power plant programs. Japan and China, on the other hand, want to resume the programs.

France claims to recycle 95% of uranium, with the other 5% going to specialized companies. All the reactors in the world produce, nevertheless, 12 thousand tons of radioactive waste. After remaining in the reactor for three years, these are converted into plutonium that can be reused. And then? After-reactions for many, for thousands of years, in a radiation spectrum that is carcinogenic to humans and animals. Melting it down in glass, embalming it in concrete, in ceramics, hiding it in shafts, that helps, to some extent. The other, simpler variant is to export to Siberia, where rust-prone tin barrels with nuclear waste simply stand in large parking lots under the open sky. But that doesn't even have to go that far: Investigations revealed that 32 ships with nuclear waste barrels on board were simply sunk in the Mediterranean.

"Experts" from business and politics currently have a new "crazy idea" on how they want to solve a problem that has so far been practically unsolved: encapsulate the nuclear waste and shoot it into the sun. They did not think about what would happen between the resulting increase in solar radioactivity and the increasing warming of the earth as a result of stronger solar rays. Do we also want to encapsulate them and send them into space, for the sake of experiences?

7 Nuclear energy for heat and electrical energy: Climate-friendly, adequate, but questionable

7.1 Application of nuclear energy in worldwide heat and electricity

From the 1950s to the present day (2023), the use of coal has been crucial in the production of electrical energy, at 38%- 40% [2]. After 2010, there was a reduction of up to 35%, but this was only partly due to the increase in the share of wind power, but more due to the use of <u>methane</u> (natural gas, LNG). **Nuclear energy** in the 1950s was already at <u>15-18%,</u> but, for reasons of various kinds, safety, *disposal of nuclear waste, but, above all, because of the huge costs during construction, decreased* to <u>10%</u>. Nevertheless, their share remains above that of *wind and solar energy* or photovoltaics. The use of nuclear energy Figure 7.1 has two fundamental advantages over other forms:

- energy production without any emission of carbon dioxide,

- absolutely stable supply of energy in the grid, without fluctuations, such as wind or photovoltaic energy generation.

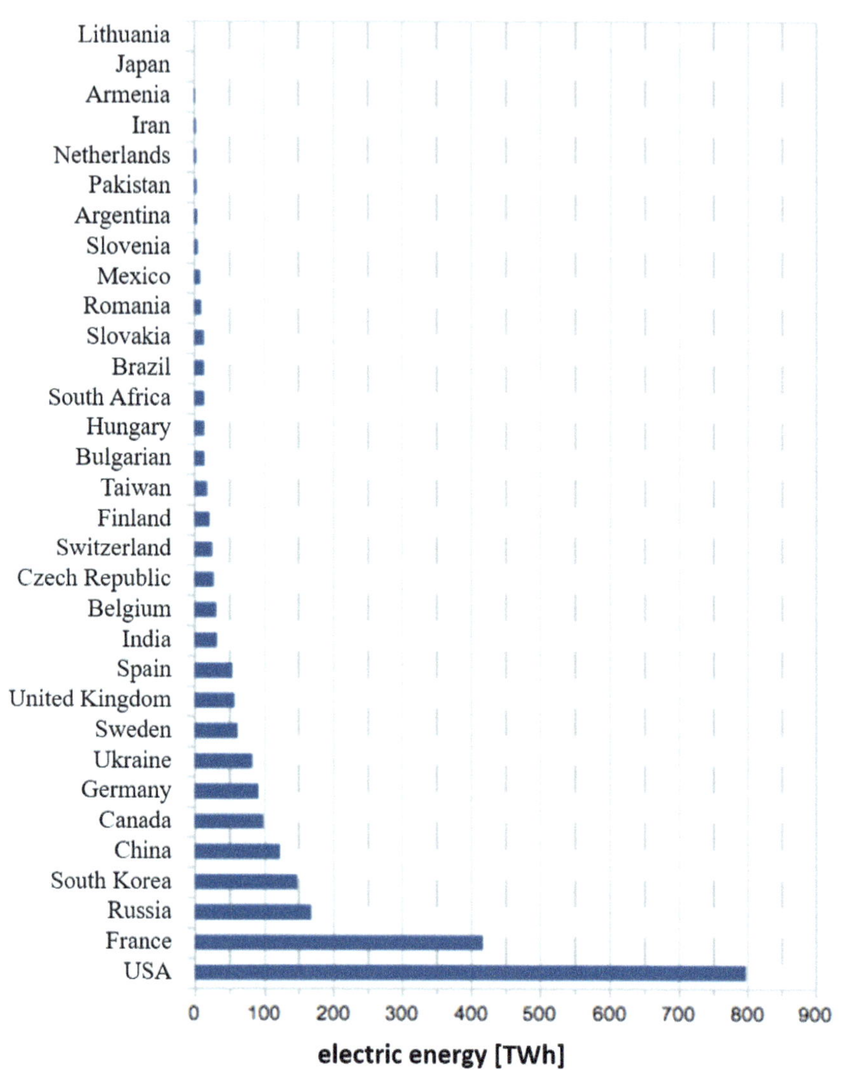

Figure 7.1 Comparison of nuclear energy production [TWh] in selected countries (*source: WNA*)

The _pros_ (*no carbon dioxide emissions at all, grid stability*), and _cons_ (*safety, disposal of nuclear waste up to the enormous costs during construction*) may be very worthy of discussion in some cases. So, at this point, let's let the facts speak for themselves.

It is obvious that nuclear energy plays a decisive role in the states with a solid economy, but at the same time with a very large demand for energy: The United States of America, France, Russia, South Korea, China, Canada.

The United States of America has (August 2022) 92 nuclear reactors in 54 nuclear (power) plants. They generated nearly 20% of the U.S. electrical energy (2021). 61 nuclear reactors operate *with pressurized water reactor* (which corresponds to the current state of the art) and 31 with *boiling water reactor.* (*The explanation of both types, pressurized water and boiling water reactor follows in chapter 7.2*)

France, second in the world in terms of the number of nuclear reactors after the United States since 2022, has *56 pressurized water reactors* distributed in the 18 nuclear power plants currently in operation [3].

These plants currently (2021) supply 61% of France's electrical energy and over 15% of the world's current output [4].

China currently (January 2023) has 55 nuclear power plants in operation and an additional 18 under construction [5].

Japan had (August 2023) 33 nuclear reactors in operation and 2 plants under construction, with 27 nuclear reactors decommissioned [6].

In the period before the Fukushima nuclear power plant disaster, 54 nuclear power plants were in operation in Japan.

South Korea operates (2022) 25 nuclear power reactors in the 4 nuclear power plants. They secure 28% of the electrical energy in South Korea. Three more nuclear power plants are currently under construction [7].

India has 22 nuclear power reactors, 7 nuclear power plants, and 8 new nuclear reactors under construction [8].

There are currently about 250 nuclear power plants in the world, which provide 10% of the country's electrical energy. In December

2022, there were a total of 439 nuclear power plants, (*With 16, in Japan, awaiting approval to resume operations*) [9], [10].

7.2   Functioning of a current nuclear power plant

Construction

At the beginning of development, between 1951 and 1957 *(USA, UK, France) there were* <u>boiling water reactors,</u> [11] consisting of a single circuit of working and cooling fluid, (*usually water).* At present, these are used by <u>pressurized water</u> reactors in almost all nuclear power reactors, as shown in Figure 7.2 They have two circuits of liquid working fluids: in the primary circuit, *the water, at a pressure of up to 160 bar,* (hence the name "pressurized water reactors" to avoid the formation of steam in *the circuit*), is used via a heat exchanger to transfer heat between the nuclear reactor as a heat source and the working fluid (*usually also water*), in the storage of a secondary circuit. The storage of the secondary circuit, which is heated with the heat from the nuclear reaction, corresponds to a "bath stove", which can be also heated with wood, gas or coal.

From this point on, in terms of the thermodynamic cycle process of the working fluid in the secondary circuit, which *converts the heat (in the steam boiler)* into work *(in a turbine or turbine series),* does not differ from the processes with steam in, for example, coal-fired power plants.

Thermodynamic Fundamentals of the Typical Steam Cycle Process in Nuclear Power Plants

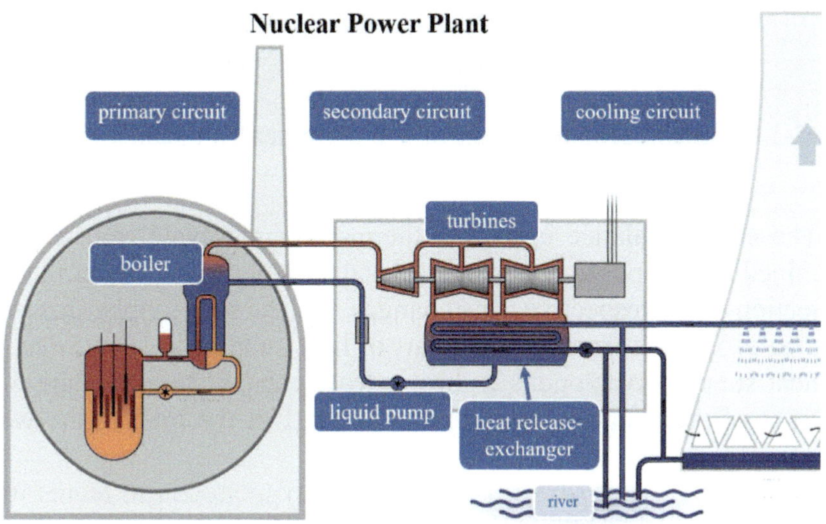

Figure 7.2 Construction of a nuclear power plant with primary and secondary circuits and modules <u>for the steam cycle process in the secondary circuit</u> (*<u>heat exchanger</u> for heat supply, <u>heat exchanger</u> for heat release, <u>turbines</u> for work and <u>water pump</u> for pumping*)

As a basis for comparison, as shown in Figure 7.3 [1] , a classical cycle process with steam – **the Clausius-Rankine process** – as well as the functional modules of a corresponding system are shown.

The thermodynamic changes of state are as follows:

1-2 <u>adiabatic</u> pressure increase of the working fluid as a liquid phase by means of a secondary water pump

2-3 <u>isobaric</u> heating of the working fluid as a liquid phase in the steam generator

3-4 <u>isobaric</u> heating of the working fluid as wet steam in the steam generator (middle part)

4-5 <u>isobaric</u> heating of the working fluid as superheated steam in the upper part of the steam generator

5-6'   <u>adiabatic,</u> irreversible expansion of the gaseous working fluid in the high-pressure turbine (5-6' ideal, isentropic expansion is shown for comparison).

6'-1   <u>adiabatic</u> condensation of the superheated steam to the liquid phase in the condenser.

The energy balance in such a thermodynamic cycle can be determined by comparing areas in the T,s diagram, namely under the projection of the respective state changes. For the heat supply, this is the area projection from state 2 (start of heat supply) to state 5 (end of heat supply). Analogous to heat release, which starts in state 6, in front of the condenser and ends in state 1 at the inlet in the water pump.

Instead of an area comparison in the <u>T, s diagram,</u> a distance comparison in the <u>h, s diagram</u> is more efficient. (*The condition for such a comparison remains that the specific heat capacity at constant pressure, which is between H and T, also remains constant, which is true in the superheated steam region*: dh= cp dT (cf. Eq. 3.14b) in [1].

The balance can be seen in the h, s diagram, by the simple route comparison in Figure 7.3/<u>h, s diagram</u>:

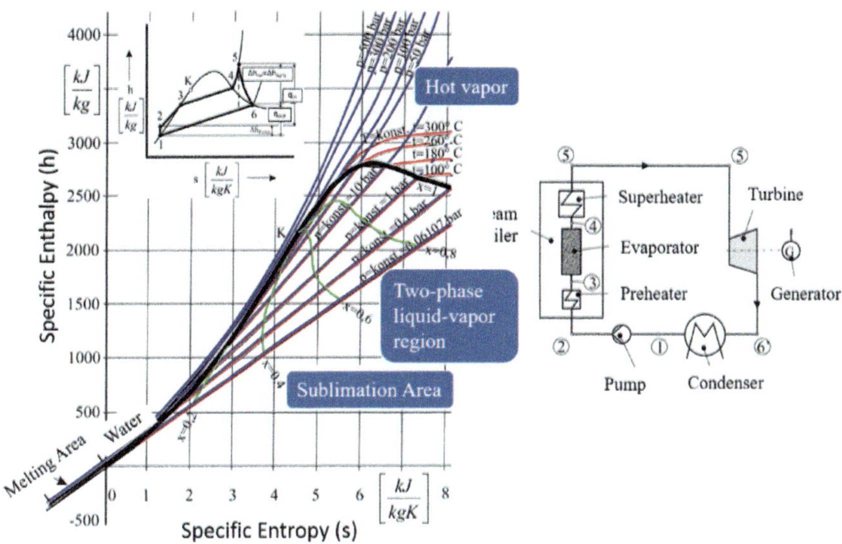

Figure 7.3 Diagram of the specific enthalpy - specific entropy of water vapor, as a basis for a Clausius-Rankine process in a nuclear power plant (small image)

$q_{in} = h_5 - h_2$	Eq. 7.1
$\lvert q_{out} \rvert = h_6 - h_1$	Eq. 7.2
$h_2 - h_1 \approx 0$, under real conditions, as work for the heat pump with $h_2 \cong h_1$ results from	
$w_K = q_{in} - \lvert q_{out} \rvert = h_5 - h_6$	Eq. 7.3
Es gilt: $h_5 - h_6 = \Delta h_{release} = \Delta h_{turbine}$ and $w_K = \Delta h_{turbine}$	Eq. 7.4
The Thermal efficiency results: $\eta_{th} = \dfrac{w_K}{q_{in}} = \dfrac{\Delta h_{turbine}}{h_5 - h_2}$	Eq. 7.4

The ratio of the distances for the work done (benefit) <u>and the heat supplied (effort) in the h, s diagram illustrates the</u> thermal efficiency for an initial assessment. Rather, it seems clear that in a thermodynamic cycle the losses are process-related.

As thermal power plants, nuclear power plants have a thermal efficiency of 33 to 40 [12].

The relatively low value **of the thermal efficiency**, similar to that of conventional power plants, with coal or gas, lies in the **low maximum temperature of the process, which,** due to the process, does not exceed 300-326°C. As a comparison, a gas flow from a **gas turbine** [13] instead of fuel jets **to heat the water in the boiler** (*will be explained in a next chapter*) brings the efficiency from over 40% to up to 63%. On the other hand, on the cold side, the restart of a thermodynamic cycle from the original state requires heat <u>release via the heat exchanger</u>.

However, the <u>T, s diagram</u> in Figure 7.3 shows how the state 6 can be moved downwards, and thus also the **minimum process temperature** can be shifted even further. For this purpose, additional low-pressure turbines have to be switched on.

7.3   Specific features of the operation of a nuclear power plant: operating time, costs, operational safety

The operating time

Beznau, in Switzerland, was commissioned in July 1969, as were the other six nuclear power plants (*such as Nine Mile Point 1 or Oyster Creek, in the USA*). In general, the lifespan of a nuclear power plant is around 60 years [14].

The duration of operation varies from country to country. In 2023, the Operating life, was on average, 31 years. (*41.2 in the USA, 36.6 in France, 8,8 in China / Sharon Wajsbrot*, [15])

Disposal of radioactive waste

According to the operators in France, 95% of uranium, even if weakened, is recycled at the nuclear power plant. Only 5% remain, of which 10% are treated by specialized firms [16].

However, there is also other information and aspects: [17]

The energy of the uranium used is much greater than the calorific value of <u>wood, coal, gas or heavy fuel oil,</u> which could also be used to heat the kettle.

<u>One kilogram</u> of uranium has as much energy as <u>12,600 liters of crude oil</u> or <u>18,900 kilograms of hard coal</u>. This is enough to generate more than 40 [MWh] of electrical energy. A fuel element remains in the reactor for about three years. After that, there is a reprocessing into plutonium, which in turn also releases a lot of energy.

And then? All of *the world's nuclear power plants currently in operation* together produce about 12,000 tons of radioactive waste per year, which also contains plutonium [17].

In addition to the problem of reactor safety during its operation, there is also the disposal and final disposal of spent radioactive components, such as fission products, and incubated trans uranium elements such as plutonium. These components are still active, although not with the same intensity as in the reactor. However, they radiate the energy generated during further fission to wavelengths in the X-ray range of the spectrum, which can be carcinogenic or even fatal to humans and animals.

These <u>post-reactions</u> take a very long time, between a few months and a few thousand years, and in the case of iodine isotopes it is even millions of years. Reprocessing would be theoretically possible, but it would "only" shorten the activity of such components to a few hundred years.

The final disposal of such "nuclear waste" remains a problem that has not really been solved worldwide.

Storage materials are not able to permanently bind or insulate such substances. They are often melted down in glass, embedded in ceramics, cast in concrete and stored in shafts, where the rock must ensure the safe containment of the radioactive materials. It would be catastrophic for water to enter such repositories, because several species of dangerous chemical reactions. Salt domes, granite and clay rock are currently being discussed as final repositories. The open storage of radioactive material in the open air is, of course, strictly prohibited by law in Western Europe.

However, the "export" of such nuclear waste to Siberia or Kyrgyzstan, where the barrels are allowed to stand in parking lots and other areas under the open sky, is not prohibited.

In 2009, the wreck of a large freighter with 120 barrels of nuclear waste on board was discovered in the Mediterranean. According to the ensuing investigations, at least another 32 ships with similar cargoes are said to have been sunk in the Mediterranean.

Scenarios for the final disposal of all the world's nuclear waste under the Antarctic ice sheet are also being developed. It doesn't get any more stupid than that! Or is it?

Disposal in space, that is the new crazy idea of "experts" from science, business, politics! Simply store them on asteroids and on other planets – perhaps even in parking lots there, like in Siberia. But the foolhardiest idea is to shoot the nuclear waste directly into the sun, so it would actually be away from our biosphere! What we then get from the dear sun as a receipt for our biosphere and for our flora and fauna, the "experts" had not come that far!

Carbon dioxide as an emission

The nuclear power plants have a very low emission of only 12 grams [CO2/kWh] [18], but according to other scientific data (CEA 2014) it was only 5.29 grams, which are caused by the construction of the nuclear power plant itself and by the extraction of uranium or other fuels.

The Costs

The cost analysis in relation to a nuclear power plant varies greatly, depending on the authors of the respective studies, or on the perspective of the observation. In some studies, the investment in the construction of the new plant is taken into account, which drives up the costs. Considering the effort required to dismantle the plant and isolate the radioactive material, the costs then explode. However, there are also numerous other studies that simply look at the current costs during the production of electrical energy.

The Rocky Mountain Institute, under A.B. Lovins, found in 2005 that considering the investment costs for the construction of the nuclear power plant and for the waste disposal and decommissioning of the reactor, the production of electrical energy by nuclear energy is not only very risky, but also, the most expensive. According to Hubert Reve's (2005) British Petroleum and Shell Oil, in view of the oil shortage, have tended to invest in renewable energies, which have a shorter time horizon and reasonable prices, also because they are subsidized by the state. A single example for orientation: in Lithuania, the construction (2007) of a nuclear power plant for 800-1600 [MW] was estimated at 2.4 to 4 billion euros [19].

Example:
for France, an RTE report was written in 2021 for the decarbonization of the country by 2050. The result is a consumption of 555 TWh/year (no significant investment in infrastructure) and 755 TWh/year (with strong industrialization). The average was 645 TWh/year, for which nuclear energy was considered up to 50%..

For an average electrical energy consumption of 360 TWh/year, the costs were calculated at 59 EUR/TWh to 57 EUR/TWh in 2040.

Accident Safety and the Function of Nuclear Power Plants in Representative Countries

After the economic crisis of 2008, and especially after the Fukushima accident, in 2011, (the <u>Fukushima nuclear disaster of</u> 2011 *refers to a series of related catastrophic accidents and major incidents: earthquakes, a seaquake that triggered a tsunami, which caused the nuclear disaster).* have led to a 4.3% reduction in the production of electricity from nuclear energy (comparison 2011-2010).

Countries such as Germany, Belgium, Switzerland and Taiwan have announced an imminent decommissioning of their nuclear power plants.

Egypt, Italy, Jordan and Thailand no longer want to enter the nuclear power plant programs.

Japan, where Fukushima is located, announced a phase-out by 2030 through the government at the time. However, after the elections of 2012, the new government has announced a resumption of the program: nine reactors have already resumed operation, six others have been given permission to restart. Cécile Asanuma-Brice (CNRS), [20], [21].

After the freezing of the programs, China decided to resume the development of nuclear power plants in 2012.

References for Chapter 7

[1]   Stan, C.: Thermodynamics for Mechanical Engineering and Vehicle Construction, 4th edition, Springer, 2020, ISBN 978-3-662-61789-2

[2]   https://ourworldindata.org/energy

[3]   https://www.iaea.org/PRIS/CountryStatistics/CountryDetails.aspx?current=FR

[4]   https://pris.iaea.org/PRIS/WorldStatistics/NuclearShareofElectricityGeneration.aspx

[5]   https://pris.iaea.org/PRIS/CountryStatistics/CountryDetails.aspx?current=CN

[6]   https://pris.iaea.org/PRIS/CountryStatistics/CountryDetails.aspx?current=JP

[7]   https://www.iaea.org/PRIS/CountryStatistics/CountryDetails.aspx?current=KR

[8]   https://www.iaea.org/PRIS/CountryStatistics/CountryDetails.aspx?current=IN

[9]   https://www.iaea.org/PRIS/CountryStatistics/ReactorDetails.aspx?current=607

[10] https://pris.iaea.org/PRIS/CountryStatistics/ReactorDetails.aspx?current=55

[11] International Atomic Energy Agency - Wikipedia https://en.wikipedia.org/wiki/International_Atomic_Energy_Agency Vienna, [PDF]*Nuclear Power Reactors in the World* [archive]

[12] https://tmuv.bayern.de/themen/reaktorsicherheit/ueberwachung/auswirkung.htm

[13] https://power.mhi.com/products/gasturbines/lineup/m701j

[14] https://www.kkg.ch/de/wissen/umwelt/energieeffizienz.html

[15] Nucléaire : comment les vieux réacteurs jouent les prolongations dans le monde entier [archive]

[16] http://www.cea.fr/energie

[17] Gerstner E: The hybrid Returns, Nature Journal Nr. 4060, Nuclear Energy, 2009

[18] https:// Centrale_nucl%C3%A9aire#GIECIII

[19] « La construction d'une nouvelle centrale nucléaire lituanienne »

[20] Huit ans après Fukushima, où en est le Japon avec l'énergie nucléaire ?

[21] ENS Lyon

Chapter 8

Jet engine flow instead of nuclear reaction: Energy for heat and electricity

In **nuclea power plants**, heat is generated by controlled nuclear fission of isotopes of uranium and plutonium, which do not cause immediate emissions <u>but do cause nuclear waste</u>. *So the goal of fission's sophisticated physical chain reactions was to superheat water.* But grandma could have done that under the bathhouse, not with fission reactions of the higher atomic physics, which she did not know, but simply with split wood. And that's what happens in power plants with coal or gas. And what if we do use gas, but not to heat the air, but to burn it right away, i.e. as exhaust gas? Instead of heat from the nuclear reaction, at 300° Celsius, just heat from the nozzle of a gas turbine, in which the huge exhaust gas flow, at up to 1200° C, almost doubles the thermal efficiency depending on the maximum temperature of the water vapor! From that point on, cause and effect appear clearer.

The <u>effect</u>: in a water cycle, superheated steam must be produced in the **water boiler** in order to make the steam turbine work.

The <u>reason</u>: it has to get hot **around the boiler**, whether it's air, exhaust gas or a different water cycle is used.

The boiler wall is the boundary between the two processes, which run completely separately. It simply does not allow mass or work to pass through, it only allows for heat transfer.

Generating heat from the nozzle of a gas turbine is much cheaper with a jet engine, be it of the type of Boeing or Airbus engines, but in terms of power and speed at the fixed operating point, which brings many advantages for jet engine construction. The advantage over diesel or gasoline engines is that the processes can be adapted in separate, specially designed rooms: compression in a compressor, combustion in a combustion chamber, relief in a gas turbine, and then exhaust gas flow to the outside via a nozzle. An engineered jet engine, which is widely used in such applications worldwide, shows the scale of such a plant: the engine (*which is installed in a combined cycle power plant with an output of 340 [MW]*) is 13 meters long, and has a circumference of about 5 meters.

The exhaust gas flow of 820 kg/second has an average temperature of 625°C.

The decisive advantage of a jet engine as a heat source, however, is the <u>unrestricted use of climate-friendly, renewable fuels</u>: **fatty acid methyl esters** (FAME **for short*) *are compounds of a fatty acid and an alcohol* (methanol). For example, a mixture of several types of FAME products, that of vegetable fats (*rapeseed oil*) or animal fats (*lard*) and alcohol (*methanol*) is already used as fuel for diesel engines (*biodiesel*). These include hydrogenated vegetable oils (HVO –or *Hydrotreated Vegetable Oils - as products of the hydrogenation of oils from plant residues or from animal fats with hydrogen),* but also ethanol *and* methanol, which can be recycled annually in nature.

Caterpillar, Mitsubishi, General Electric Power Ansaldo and, above all, Siemens are well-known pioneers of this process.

8 Jet engine flow instead of nuclear reaction: energy for heat and electricity

8.1 Structure and operation of a gas and steam turbine Combined Cycle Gas Turbine (CCGT) power plant

In **nuclear power plants**, nuclear fission is used to generate heat in **the primary circuit,** which is transferred from a circuit of water **(primary circuit)** via a heat exchanger to a water/steam cycle process **(secondary circuit)**.

The primary circuit with nuclear reaction can be replaced by the hot exhaust gas flow from a jet engine. This has the advantage of significant cost savings, as well as an increase in thermal efficiency by increasing the maximum temperature from about 40% to almost 63%. This is explained as follows:

Instead of heat from the nuclear reaction (300-320°C), heat comes from the exhaust gas flow of a gas turbine (up to 1200°C *at the turbine outlet*). The receiver is, in both cases, a heat exchanger that is transferred by *convection* to the working fluid in the secondary circuit. From there, in principle, *a similar* Clausius-Rankine steam cycle process *runs (by means of the steam),* for which similar functional modules are used in the secondary circuit of the combined-cycle power plants as well as in the nuclear power plants.

This also ensures short start-up times for the combined-cycle power plant and the associated ability to react quickly to power grid requirements.

© The Author(s), under exclusive license to
Springer-Verlag GmbH, DE, part of Springer Nature 2024
C. Stan, *Energy Scenarios for the Future,*
https://doi.org/10.1007/978-3-662-69687-3_8

In addition to the separate gas turbine and steam turbine turbo sets **(multi-shaft version, possibly with 2 power generators),** single-shaft versions are used in the higher power segment **(a gas turbine and a steam turbine** drive a power generator *simultaneously on a common shaft train).*

The gas turbine is operated with conventional fuels in liquid or gas form. It is currently not possible to use **coal dust** as fuel in the gas turbine because coal combustion produces ash that would destroy the blades of the gas turbine. Separating the ash from the hot gas stream is technically very complex.

A variant of this is the *coal-fired combined-cycle* power plant: a coal-fired steam power plant in combination with a jet engine that runs on **natural gas**. According to the gas turbine of the jet engine, the exhaust gases have a high temperature and about 15 % oxygen content, which can be used as supply air for a coal-fired steam generator, which in turn supplies the superheated steam for a steam turbine in the steam cycle process.

By dissipating the energy/enthalpy at a certain steam turbine pressure stage, the thermal energy of the water vapor can still be used for **heat**.

Other manufacturer designations include *Steam and Gas (STAG)* from General Electric or *Kombi (KA)* from Alstom. This is referred to as a Combined Cycle *Power Plant (CCPP) or Combined Cycle Gas Turbine (CCGT).*

8.2 Thermodynamic Fundamentals of the Steam Cycle Process in the Gas-and-Steam Cycle Combination

Secondary circuit in nuclear power plants or in gas and steam cycle combination

In contrast to a piston engine, the jet engine is regarded as an open machine, in which the four similar state changes (heat *supply/heat release, compression/expansion) take place* with a constant mass flow of the gaseous working fluid (*ideally air, then exhaust gas mixture after combustion*). Compression and expansion are similar to those of the piston engine (*ideally without heat exchange with the environment*, i.e. isentropic). However, due to the mass flow of the working fluid, both the heat supply q **23** and the heat release **q 41** take place at the same pressure (isobaric).

The heat supply is much simpler than in a piston engine: injectors, as in fuel direct injection in piston engines, which open and close in microseconds, are superfluous in the case of a constant flow of compressed air. Instead, open (swirl) nozzles can be used. Hence the jet engine's ability to be compatible with almost any fuel, from solid (*when it burns without particles that can damage the turbine blade*) to liquid and gaseous. The isobaric heat release **q 41** in the jet engine has the significant advantage of a much greater discharge compared to a piston engine (*heat dissipation at the same volume, limited by the geometrically limited piston stroke*), which greatly benefits the thermal efficiency.

Figure 8.1 (top) shows the heat discharge q 41 (q$_{out}$) of the thermodynamic cycle of a jet engine in the T,s diagram. This heat as heat absorption q23 (**q 25**) in the secondary circuit (below) following the blue arrow. It should be emphasized that, for the secondary circuit, it is irrelevant from which source the absorbed heat (**q 25**) originates, whether it is from the primary circuit of water heated in the nuclear reactor or exhaust gas produced in the gas turbine.

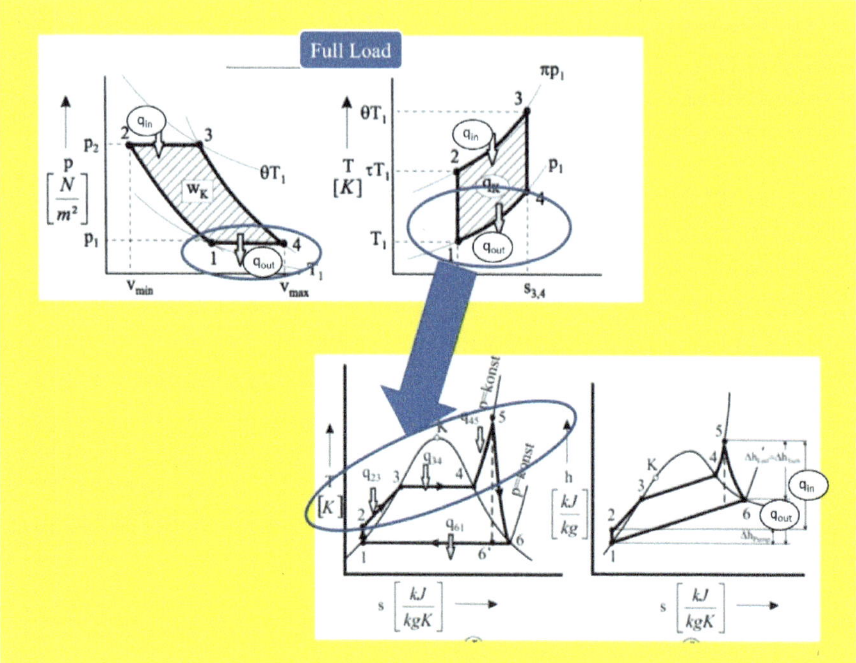

Figure 8.1 Heat dissipation q 41 (q$_{out}$) from the thermodynamic cycle of the gas turbine (top) in the T, s Diagram and heat absorption q 25 (q$_{in}$) in secondary circuit (bottom)

For the steam cycle process into the secondary circuit, the thermodynamic relationships are shown in Chap. 7.2, with the mentions:

Figure 8.2 shows [1] a cycle process with water-steam – the Clausius-Rankine process – in the T,s diagram, as well as the functional modules of a corresponding plant.

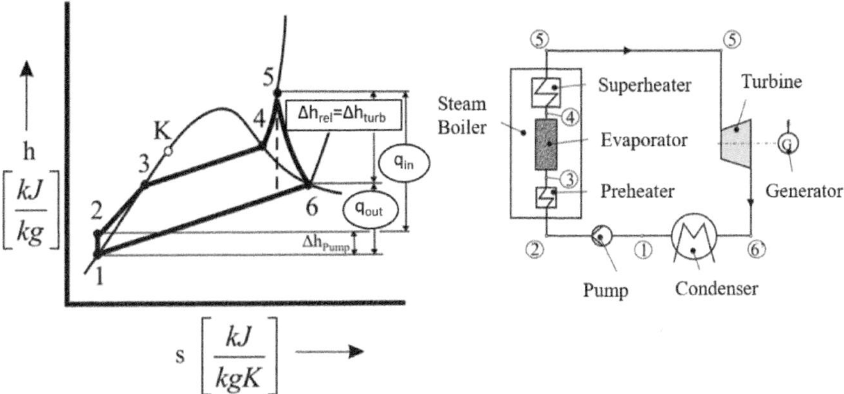

Figure 8.2 Clausius-Rankine thermodynamic cycle for water vapor in T, s- diagram and function modules (schematic) in order to perform the above transformation of state

There are following transformation of state changes:

-   1-2 adiabatic pressure increase of the working fluid as a liquid phase by means of a pump

-   2-3 Isobaric heating of the working fluid as a liquid phase in the steam generator

-   3-4 Isobaric heating of the working fluid as wet steam in the steam generator (middle part)

-   4-5 Isobaric heating of the working fluid as superheated steam in the upper part of the steam generator

-   5-6 adiabatic, irreversible discharge of the gaseous working fluid in the high-pressure turbine (5-6 ideal, isentropic discharge is shown for comparison).

-   6-1 Condensation of the superheated steam to the liquid phase in the condenser..

The energy balance in such a thermodynamic cycle can be determined by comparing areas in the T, s diagram, namely under the projection of the respective state changes. For the heat supply, this is the area projection from state 2 (start of heat supply) to state 5 (end of heat supply). Analogous, this ca be made to heat release, which starts in state 6, in front of the condenser and ends in state 1 at the inlet in the water pump.

However, instead of an area comparison in the T, s diagram, a distance comparison in the h, s diagram is more efficient, as mentioned in Chapter 7. (cf. Eq. 3.14b) in [1].

The balance can be seen in the h,s diagram, by the simple route comparison in Figure 7.3/h.s diagram:

$q_{in} = h_5 - h_2$	Eq. 7.1
$\left\lvert q_{out} \right\rvert = h_6 - h_1$	Eq. 7.2
$h_2 - h_1 \approx 0$, under real conditions, as work for the heat pump results: $h_2 \cong h_1$ from:	
$w_K = q_{in} - \left\lvert q_{out} \right\rvert = h_5 - h_6$	Eq. 7.3
It is: $h_5 - h_6 = \Delta h_{release} = \Delta h_{turbine}$ und $w_K = \Delta h_{turbine}$	Eq. 7.4
The thermal efficiency results as: $\eta_{th} = \dfrac{w_K}{q_{in}} = \dfrac{\Delta h_{turbine}}{h_5 - h_2}$	Eq. 7.4

The ratio of the distances for the <u>work done</u> (benefit) <u>and the heat supplied (effort)</u> in the h, s diagram illustrates the thermal efficiency for an initial assessment. Rather, it seems clear that in a thermodynamic cycle the losses are process-related.

As thermal power plants, nuclear power plants have a thermal efficiency of 33 to 40% [2].

The relatively low value **of the thermal efficiency**, similar to that of conventional power plants, with coal or gas, is **due to the low maximum temperature of the process,** which, due to the process, does not exceed 300-326°C in the nuclear reactor. For comparison, **a gas flow from a gas turbine** [3]:

Instead of using fuel jets to heat the water in the boiler *(explained in a next chapter)*, brings the efficiency up to 63%. On the other hand, _on the cold side_, the restart of a thermodynamic cycle from the original state entails a heat _release_ *(in this case via the heat exchanger).*

In the T, s diagram in Figure 7.3 , however, it was possible to see how the state 6 can be moved downwards, and thus also how the **minimum process temperature** can be shifted even further. For this purpose, additional _low-pressure turbines have_ to be switched on..

Figure 8.3 shows a comparison of the processes in a _nuclear power plant_ (top) and in _a combined-cycle power plant_ (bottom). The respective connection of the two different working fluids (water in the primary circuit of the nuclear reactor, top) and gas (in the gas turbine, bottom) with the **secondary circuit** is shown in this sketch *as a mass-tight heat exchanger* (blue arrow) schematically as an arrow. _The secondary circuits of both processes are basically the same._

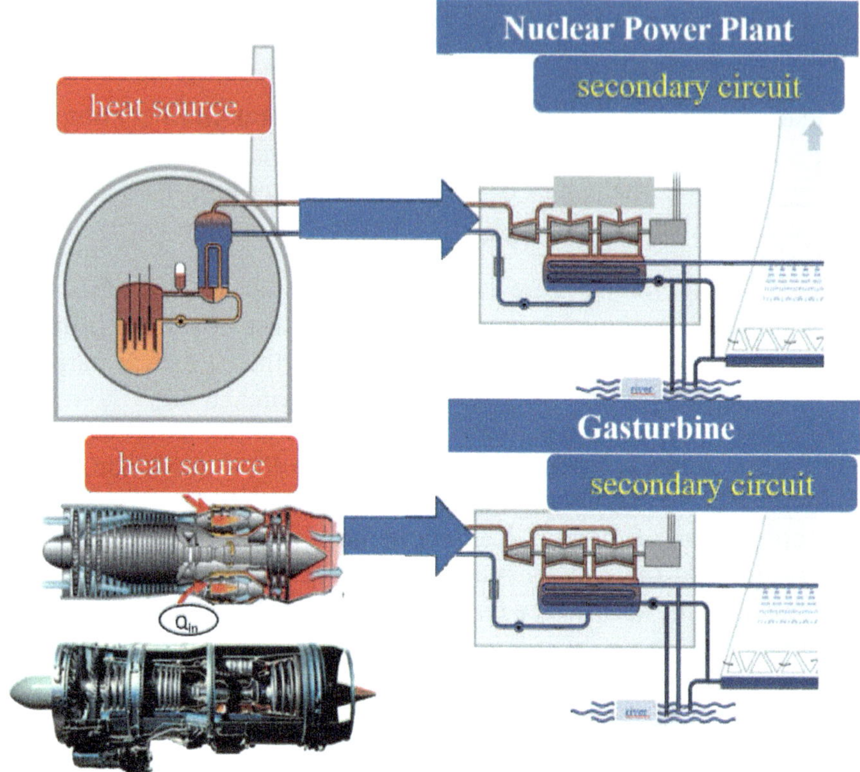

Figure 8.3 Comparison of the processes in <u>a nuclear power plant (top) and in</u> a gas-steam power plant (bottom). The respective connection of the two different working fluids (water in the primary circuit of the nuclear reactor, top) and gas (in the gas turbine, bottom) with the **secondary circuit** is to be regarded *as a mass-tight heat exchanger (*blue arrow, symbolic). <u>The secondary circuits of both processes are basically the same</u>

Flow from gas turbine as a heat source instead of reaction heat in the primary circuit of a nuclear power plant

After all, a gas turbine or a jet engine is used in airplanes, as turbofans or turbojets, in all kinds of large passenger aircraft and fast military machines of all kinds. The application in a power plant, also if indirect, but very advantageous, um Generating electricity or heat seems astonishing at first glance. The mode of action is as follows:

A turbomachine, as mentioned above when performing the process, generally contains a *compressor, a combustion chamber and a turbine*. The <u>compressor</u> is directly (mechanically) coupled to a <u>turbine module</u>. This turbine module takes the part of the supplied energy that is necessary to drive the compressor and converts it into kinetic energy. Otherwise, the remaining energy is converted for the <u>actual propulsion</u>, either via a <u>gearbox</u> or via a <u>nozzle</u> to produce the reaction force. Another possibility is to use the remaining energy in a supply line, immediately after the turbine, where it has a *sufficient energy/enthalpy*, expressed in the high temperature of the flow, as shown in the T,s diagram in Figure 8.4.

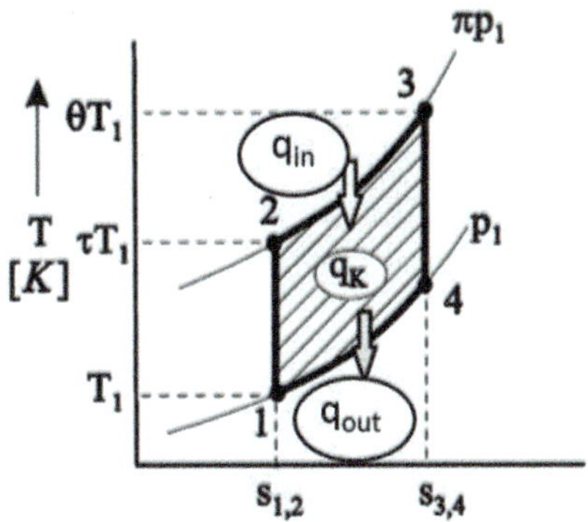

Figure 8.4 jet engine and T, s diagram for its thermodynamic cycle (working medium: Gas)

As an example of an engineered jet engine, which is widely used worldwide, shows the magnitude of such a plant: the engine, (which is *installed in a combined cycle power plant with an output of 340 [MW]*) is 13 meters long, and has a circumference of about 5 meters. The exhaust gas flow of 820 kg/second has an average temperature of 625°C.

8.3 Application Perspectives and Functional Features

Development of thermal efficiencies (examples)

- In May 2011, power plant unit 4 in Irsching achieved an efficiency of 60.75 % in trial operation, thus becoming the new world record holder [4].

- In January 2016, the Lausward power plant in Düsseldorf, Germany, achieved an efficiency of 61.5% during test runs using a Siemens SGT5-8000H gas turbine. With maximum heat extraction, more than 85% fuel efficiency is achieved and carbon dioxide emissions of only approx. 230 g/kWh [5] are achieved.

- In April 2016, the Électricité de France in Bouchain, which uses General Electric's 9HA gas turbine, entered the Guinness Book of World Records as the most efficient combined cycle power plant with a combined efficiency of 61.4% [6].

- In March 2018, the Nagoya power plant, powered by General Electric's 7HA gas turbine, entered the Guinness Book of World Records as the most efficient combined cycle power plant with an efficiency of 63.08% [7].

- In 2018, General Electric offered an 826 MW gas turbine that is expected to achieve an efficiency of more than 64% in the combined cycle; a combination efficiency of 65% was targeted for the early 2020s [8], [9].

Costs

Combined-cycle power plants can be built comparatively quickly and inexpensively. The construction time and the investment costs are only about half of that of a coal-fired power plant of the same capacity. However, in the base load range with 8000 operating hours per year, coal-fired power plants can compensate for the higher construction costs due to lower fuel costs. The more flexible combined-cycle power plants are therefore primarily used in the peak and medium-load range with 4000 operating hours per year [10].

FAME (*fatty acid methyl ester*) as a climate-friendly fuel for gas turbines in combined-cycle power plants

Fatty Acid Methyl Esters **(FAME*) are compounds of a fatty acid and an alcohol** (methanol). For example, a mixture of several types of FAME products, that of vegetable fats (*rapeseed oil*) or animal fats (*lard*) and alcohol (*methanol*) is already used as fuel for diesel engines (*biodiesel*).

Fatty acid methyl esters (FAME) are produced by transesterification of fats or oils (triglycerides) with methanol. In this process, glycerin is exchanged for methanol. Glycerin and FAME are produced as reaction products.

Today, fatty acid methyl esters are mainly used for the production of biodiesel and can be used as a pure fuel as well as in any mixture with conventional diesel fuel. As a proportioning component for diesel, the fatty acid methyl ester must meet certain, precisely defined quality parameters.

FAME have a significantly lower viscosity than untreated vegetable oil, so it can be used as a substitute for mineral diesel fuel without the need to adapt the *diesel engine or jet engine*. However, fuel system components that come into contact with fuel, such as hoses and seals, must be resistant to the methyl ester and must be made of polytetrafluoroethylene (Teflon) or Fluor rubber (e.g. Viton). FAME can also be used to produce fatty alcohols and fatty amines, which are used, for example, to produce surfactants and emulsifiers. The most common fatty acid methyl esters are soybean oil methyl ester (SME; mainly in North and South America, also imported into Europe), rapeseed methyl ester (RME; mainly in Central Europe), palm oil methyl ester (PME) and fat methyl ester (FME) derived from animal fats.

Prospects for use of the combined-cycle power plant as an efficient and climate-friendly power plant of the future

The *companies Caterpillar, Mitsubishi, GE Power, Ansaldo* and *Siemens* are increasingly operating combined cycle power plants worldwide. In Irsching, near Ingolstadt, for example, a plant with 578 [MW] and a thermal efficiency of 60.75% is in operation. The largest works of this kind in Germany are Knapsack I and II. Together, they create 1,230 [MW].

For comparison, the largest power plant in the world, the Three Gorges Dam in China (*as a hydroelectric power plant*) has a capacity of 22,500 [MW].

Hydrogenated vegetable oils (HVO – *Hydrotreated* **V**egetable **O**ils - as products of the hydrogenation of oils from plant residues or from animal fats with hydrogen), *but also ethanol and methanol and climate-friendly hydrogen* that can be recycled annually in nature are a significant contribution to the production of heat and electrical energy efficiently, cost-effectively and in significant quantities, but above all, without impact on the climate.

References for Chapter 8

[1]   Stan, C.: Thermodynamics for Mechanical Engineering and Vehicle Construction, 4th edition, Springer, 2020, ISBN 978-3-662-61789-2

[2]   https://tmuv.bayern.de/themen/reaktorsicherheit/ueberwachung/auswirkung.htm

[3]   https://power.mhi.com/products/gasturbines/lineup/m701j

[4]   Das leistungsfähigste Kraftwerk der Welt. In: Bild der Wissenschaft. 20. May 2011, retrieved 9 September 2019

[5]   Düsseldorf: Kraftwerk bricht zahlreiche Weltrekorde (Memento vom 28. Januar 2016 im Internet Archive). In: Zeitung für kommunale Wirtschaft, 28. Januar 2016. retrieved 28. January 2016.

[6]   Most efficient combined cycle power plant.

[7]   GE-Powered Plant Awarded World Record Efficiency by Guinness. Power Engineering, 27. März 2018

[8]   https://de.wikipedia.org/wiki/Gas-und-Dampf-Kombikraftwerk - cite_note-8

[9]   HA technology now available at industry-first 64 percent efficiency. GE Power, 4. December 2017

[10]   https://www.guinnessworldrecords.com/world-records/431420-most-efficient-combined-cycle-power-plant

Chapter 9

Heat and electrical energy from waste and biomass to reduce greenhouse gas emissions

We produce so much waste: **1.3 kilograms per day,** which is **475 kilograms per** inhabitant per year, the Germans are somewhat more economical with 455 kilograms.

The remaining energy in the waste *(rubble, waste wood, paper and cardboard)* can be converted into heat by incineration, some of which can be used as work. Reduction is not avoidance of greenhouse gas emissions, but at least it is a contribution in the right direction.

In 2200 waste incineration plants worldwide, **255 tons of waste were incinerated annually.** An increase to **430 million tons of waste** is expected soon. The calorific value of combustion with air is admittedly only a quarter of that of Combustion of diesel fuel and gasoline results, but heat is heat. A two-stage combustion process is generally used, as in earlier vortex chamber or pre-chamber *(like in Diesel engines)*: combustion in the absence of oxygen, then completely, resulting in little carbon monoxide and nitrogen oxides, and so, in carbon dioxide, proportional to the amount of waste and air involved. And so, a water boiler is heated, as in a nuclear reactor, as with coal and gas, or even with the jet engine. Due to batteries, PVC, exhaust gas cleaning is particularly important in this case.

800 thousand tons of hard coal, but also 650 thousand tons of residual waste, are burned in the *Munich-Nord combined heat and power plant*, which produces 900 megawatts of heat and 411 megawatts of electricity. By way of comparison, in *Bolzano/South Tyrol*, where *52% of the region's waste is recycled, 59 megawatts of* heat and 15 megawatts of electrical energy are generated from 130 thousand tons of residual waste. Such processes in which heat and electricity are produced simultaneously *(combined heat and power) have recently led to modular plants, commonly known as* Combined Heat and Power plants **(CHP).** Biogas is often used instead of *waste.* This also

changes the configuration of the system itself: diesel engine instead of boiler and steam turbine, engine cooling water instead of boiler. By using the stationary diesel as a heat source via cooling water, the thermal efficiency of the diesel engine increases to 80 to 90%. However, it will not only be very efficient, but also climate-friendly: cow *manure* or *pig manure,* which can be found in every farmer and village worldwide, is sufficient for biogas, as concrete examples prove. And on the horizon is the electrolytically generated hydrogen generated by wind and sun!

9   Heat and electrical energy from waste and biomass to reduce greenhouse gas emissions

Waste incineration process

The incineration of the atmospherically combustible parts of residual waste serves to obtain the energy still contained in such waste (construction waste, waste *wood, glass, small scrap, paper and cardboard, packaging)* that would otherwise not be used and at the same time to compact the amount of waste. Although this is a significant *reduction*, it *is not an avoidance* of greenhouse gas emissions.

There were 2200 waste incinerators worldwide (as of 2015) in which 255 million tons of waste were incinerated. By 2025, it is expected to increase to 2750 plants for 430 million tons of residual waste [3].

Waste, regardless of the varieties listed above, is the fraction that can burn with oxygen from the air at ambient pressure. The calorific value of such waste is 25% of the usual values for gasoline and diesel fuel [1]. Up to 0.36 [kWh] of electrical energy can be obtained from one kilogram of wet waste, considering the process stages and the associated efficiencies [4].

For example, in a waste incineration plant, after drying the waste at about 100 °C, degassing is carried out at 250-900 °C and then incineration under oxygen deficiency at 800-1150 °C, resulting in carbon monoxide and unburned hydrocarbons with low nitrogen oxide emissions [1].

In a further stage of the firing process, air is added again, whereby the intermediate products are completely burned into carbon dioxide and water. This two-stage combustion process (*similar to that in earlier diesel engines with pre- and vortex chambers*) ultimately serves a complete combustion with a lot of carbon dioxide and as little carbon monoxide and nitrogen oxides as possible [4].

The resulting flue gas transfers the heat to the heating surfaces of a steam boiler for hot water.

When the waste is incinerated, it is generally not known which substances contained in it are included in the reaction and in what quantities at a given time. For example, PVC, batteries, electronic components and paints are critical, which can also produce hydrochloric acid (, hydrogen fluoride, as well as mercury and dusts containing heavy metals. For this reason, exhaust gas purification is particularly important.

Each inhabitant of Europe produces an average of ***475 kilograms of waste*** annually which is **1.3 kilograms** per day! [5]

In the Munich North combined heat and power plant, ***800,000 tons of hard coal*** are fired annually in a first plant module, and 650,000 tons ***of residual waste are incinerated*** in another two modules [6]

The combustion of these energy sources generates **900 [MW] of heat and 411 [MW] of electrical energy**. The annual carbon dioxide emissions of the entire plant amount to around 3 million tons per year (2015). By way of comparison, Thyssen Krupp's Duisburg steel mill produces 8 million tons of CO2 annually.

One of the world's most modern combined heat and power plants with waste incineration is located in Bolzano/South Tyrol, Italy. South Tyrol has one of the strictest environmental laws in Europe, from the collection to the recovery of all types of waste: 52% of the region's waste is recycled, 44% is incinerated, only 4% is stored. The waste incineration plant incinerates *130,000 tons* of residual waste annually, producing **59** [MW] of heat and **15** [MW] of electrical energy [7].

All pollutants emitted are far below the particularly low permissible limits. *Dioxin 1% of the standard,* nitrogen oxides *15% of the standard,* particles 8% of the standard. The exhaust gas consists practically only of **carbon dioxide** and water vapor. Figure 9.1 shows a schematic representation of the combined heat and power plant with waste incineration in Bolzano, Italy.

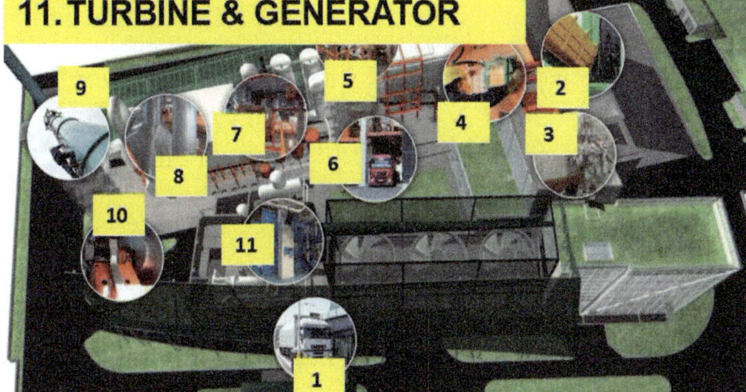

Figure 9.1 Combined heat and power plant with waste incineration in Bolzano/ Italy – functional diagram (*Source: ECO Center AG, Bolzano*)

The regional recycling of waste not only solves the problem of overflowing landfills, but also contributes to the supply of heat and electricity, with a considerable share, in addition to the central supply networks. However, it is first and foremost a significant contribution to reducing carbon dioxide pollution.

Combined heat and power plants based on combined heat and power

A combined heat and power plant (**CHP** is a modular plant for the production of electrical energy and heat, which is often operated at the point of heat consumption, on the basis of combined heat and power (**CHP**), in which mechanical energy is usually converted directly into electricity. If the heat is used for heating purposes as local or district heating or for production processes as process heat, it is referred to as **an industrial power plant**.

A frequent use of combined heat and power plants based on combined heat and power is with biogas combustion instead of waste incineration.

The use of biogas within a biogas plant, by means of internal combustion engines, which act as generator drives to generate electrical energy, is particularly advantageous. In Germany, there are 9500 such decentralized plants (2019), in other countries this type of use of biogas is increasing.

Example:
*In the small biogas plant in a rural Eastern European region, in Transylvania, 370 kilowatt-hours [kWh] of electrical energy are produced daily from 55 tons of **cow dung** from a single farm. This energy would be enough to fully charge the 32.3 kWh batteries of 11 compact electric Volkswagens [2].*

Biogas consists of 50-75% methane. Methane as an energy source has about the same calorific value as gasoline and diesel fuel [1]. It is therefore also used as a fuel or fuel in *heating systems, combined heat and power plants and heat engines.*

Biogas is produced by the fermentation of biomass of all kinds – organic waste (*food waste, lawn clippings), manure, manure, plant residues* or specifically cultivated *energy crops.*

For example, pig manure has a biogas yield of 60 m³/ton with 60% methane content, chicken manure 80 m³/ton with 52% methane content, *biowaste 100 m3/ton with 61%* methane content.

Biogas is the explosive product of the decomposition of organic components in the biomass by microorganisms, in the absence of oxygen. During this process, the **carbohydrates,** proteins and fats contained are mainly converted into methane and carbon dioxide.

The methane from biogas can be mixed with natural gas in any proportions due to the same properties. Both can be used separately or in variable mixtures in combustion plants and in heat engines for stationary or mobile use as fuel/fuel. This is a good prerequisite for the rapid transition from fossil to renewable fuels.

In addition *to biogas* and natural gas, there is another one with considerable future potential in terms of climate protection, as long as it is produced without carbon dioxide emissions (*for example by water electrolysis*): *hydrogen,* which can be mixed up to 30% with the other two [8].

Heat and electricity with biogas by means of internal combustion engines

The path to biomethane runs through several process stages, from hydrolysis and acidogenesis to acetogenesis and methanogenesis. In addition to methane, this also produces water, but also a proportion of carbon dioxide.

Biogas, currently still mixed with natural gas from the existing natural gas networks, is the most commonly used fuel in combined heat and power plants in which electrical energy and useful heat are generated. Gasoline and diesel engines, gas and steam turbines, Stirling engines and steam power plants are used as heat engines to convert fuel energy into mechanical work for power generators and into heat.

The advantage of combined heat and power plants, as shown in Figure 9.2, over separate plants for the production of heat and electrical energy lies in the more efficient use of fuel energy.

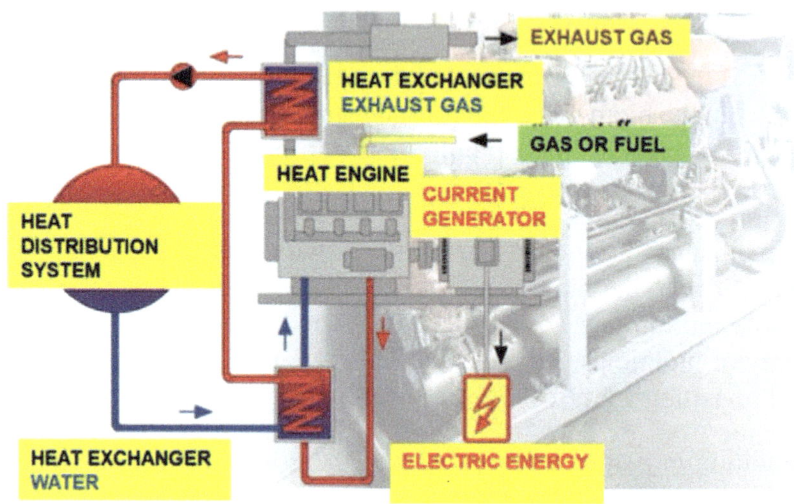

Figure 9.2 Combined heat and power plant based on combined heat and power: the piston engines operate practically stationary, cooling water and exhaust gas are fed into a heating network

In power plants for the sole generation of electrical energy by means of a heat engine, which is often a classic combustion engine, about 30% of the heat supplied for engine cooling and an equal percentage of the exhaust heat is generally released into the environment unused.

In power plants that generate heat in addition to electricity, these two proportions are used for heating purposes or for other heating applications, increasing the overall thermal efficiency of such an internal combustion engine to 80% to 90%.

In two new plants of this type, which were commissioned in Chemnitz, Germany [9] (2020), 5 and 7 large gas engines are used respectively to generate 150 [MW] electrical **and 130 [MW] thermal** power: In addition, three new boilers will be used to supply heat to the city, which will also be fired with natural gas/biogas and will provide a total thermal output of **100** [MW].

The combined heat and power plants based on heat and power and boilers with the same fuel, natural gas/biogas, is currently the most modern form of supply of electricity and heat, thus achieving a reduction in carbon dioxide emissions of up to 40% compared to traditional coal-fired power plants.

References for Chapter 9

[1]   Stan, C.: Thermodynamics for Mechanical Engineering and Vehicle Construction, 4th edition, Springer, 2020, ISBN 978-3-662-61789-2

[2]   Stan, C.: Energy versus carbon dioxide, Springer, 2021, ISBN 978-3-662-62705-1

[3]   https://www.berlin-recycling.de/blog/impulse/525

[4]   https://www.swd-ag.de/magazin/muellverbrennung/

[5]   Markus Gleis: *Thermal Waste Treatment*. In: Peter Kurth, Anno Oexle, Martin Faulstich (eds.): *Praxishandbuch der Kreislauf- und Rohstoffwirtschaft*. Springer Vieweg, Wiesbaden 2018, ISBN 978-3-658-17044-8, S. 621

[6]   https://www.swm.de/energiewende/kraft-waerme-kopplung

[7]   https://www.eco-center.it/de/aktivitaeten-dienstleistungen/umwelt/.html

[8]   https://ingenieur.de/technik/fachbereiche/energie/deutsche-gasleitungen-sind-fuer-wasserstoff-geeignet/

[9]   https://www.eins.de/ueber-eins/infrastruktur/erzeugung/heizkraftwerk-chemnitz

Chapter 10

Alternative fuels for environmentally friendly heating and working machines

A fuel, be it hydrogen, hydrocarbon or alcohol, can be used as a double base: to generate heat, *and thus* work in a diesel engine or **gasoline engine**; to generate **electricity** in *a fuel cell* as a result of a proton exchange.

When carbon **is present in the structure of an energy source, whether fossil** or renewable, it *results in any energy conversion, by combustion or by any other chemical reaction,* (heat or electrical energy) and carbon dioxide. But there is a difference between *fossil and renewable energy sources:* the partial recycling of carbon dioxide *produced by renewable energy sources, is given by carbon dioxide which is absorbed from the atmosphere within the photosynthetic plant nutrition cycle*. In this context, ethanol **from plants** and **biowaste** has considerable potential for the future of combustion in **ethanol vehicles**: these numbered around fifty million units **in 2017**. At the same time, **methanol is produced from industrial carbon dioxide emissions**, of which Germany has no less than 800 million tons, and also from **waste incineration**. The hydrogen required for synthesis in methanol is produced electrolytically in exemplary applications, using electricity from *photovoltaics* and *wind energy.*

Fuel properties are as important as *resources, environmental behavior and technical basis:* the mass *fractions* of carbon, hydrogen *and* oxygen *are decisive for the mass of the combustion air* required but also for the mass *of carbon dioxide emissions.* The *density* determines the tank volume with comparable stored energy. The energy developed per kilogram *of fuel (fuel calorific value) is unsurpassed for hydrogen, but the energy of the chemically determined* **mixture hydrogen/air (mixture calorific value)** during combustion is almost the same for all mixtures of air with gasoline, ethanol or hydrogen. *This means that switching to an alternative fuel in a combustion plant is not expected to result in a spectacular change in the specific work [kJ/kg mixture].*

The knock resistance (expressed by the <u>octane number</u>) and the ignition (expressed by the <u>cetane number</u>). The **evaporation properties** (*for example, how <u>quickly</u> and <u>how much heat</u> it extracts*) are something you want to ignore at first. However, this has a decisive influence on the combustion process: fine droplets, almost steam and the lowering of the combustion temperature by heat extraction for evaporation are the best prerequisites for intensive combustion with a maximum heat development, with very little pollutant content.

There are a large number **of renewable fuels** (methanol, ethanol, hydrogen, dimethyl ether) *whose production is already efficient and on a convincing scale* (ethanol from simple distillation or from waste in bioreactors, by means of cyanobacteria, methanol synthesis from carbon dioxide and hydrogen).

In addition, fuels such as **synfuel** or **"designer fuel"** are already produced in promising quantities by proven methods of process engineering *(refinery, distillery, gas synthesis, pyrolysis, electrolysis, Carbo-V/Fischer-Tropsch process, hydrogenation process, pyrolysis process).*

But that's not all when it comes to climate-friendly, renewable fuels: **FAME** (*fatty acid methyl ester*) consists of compounds of a *fatty acid* and *methanol*. To produce <u>*fatty acid methyl ester*</u> (FAME), a trivalent alcohol glycerol is exchanged for methanol. Glycerin and FAME are produced as reaction products.

Hydrogenated vegetable oils are produced in existing refineries together with other fats and mineral oil components as well as in our own vegetable oil plants.

There is immense potential in the countless combinations of thermochemical and electrochemical processes: It opens the way to the controlled construction of molecular structures of climate-friendly fuels that do not cause carbon dioxide emissions during combustion.

It is not the "combustion engine" that needs to be replaced, but what it has to burn.

10 Alternative fuels for environmentally friendly heating and working machines

10.1 Climate-friendly energy sources: resources, processes, utilization

The chemical energy of any fuel, _hydrocarbon_ or _alcohol_ (CmHnOp) can basically be converted into two forms:

In **heat,** as a precursor to another form of energy - work**,** for use in **internal combustion engines** and further for propulsion or for on-board power generation.

In the same chain of chemical energy-heat-work-electricity, heat and power are used in thermal power plants, combined heat and power plants **(CHP), on the basis of** combined heat and power, **or in** industrial power plants.

In **electrical energy** as the basis of _the proton exchange hydrogen-oxygen in the fuel cell_. The energy source for the hydrogen in such a reaction can be, in turn, a _carbon_ or an _alcohol_.

Figure 10.1 shows an overview of _energy sources and the processes used for them_. On the lower side of the picture are the fossil fuels, on the upper side the renewable energy sources.

© The Author(s), under exclusive license to
Springer-Verlag GmbH, DE, part of Springer Nature 2024
C. Stan, _Energy Scenarios for the Future_,
https://doi.org/10.1007/978-3-662-69687-3_10

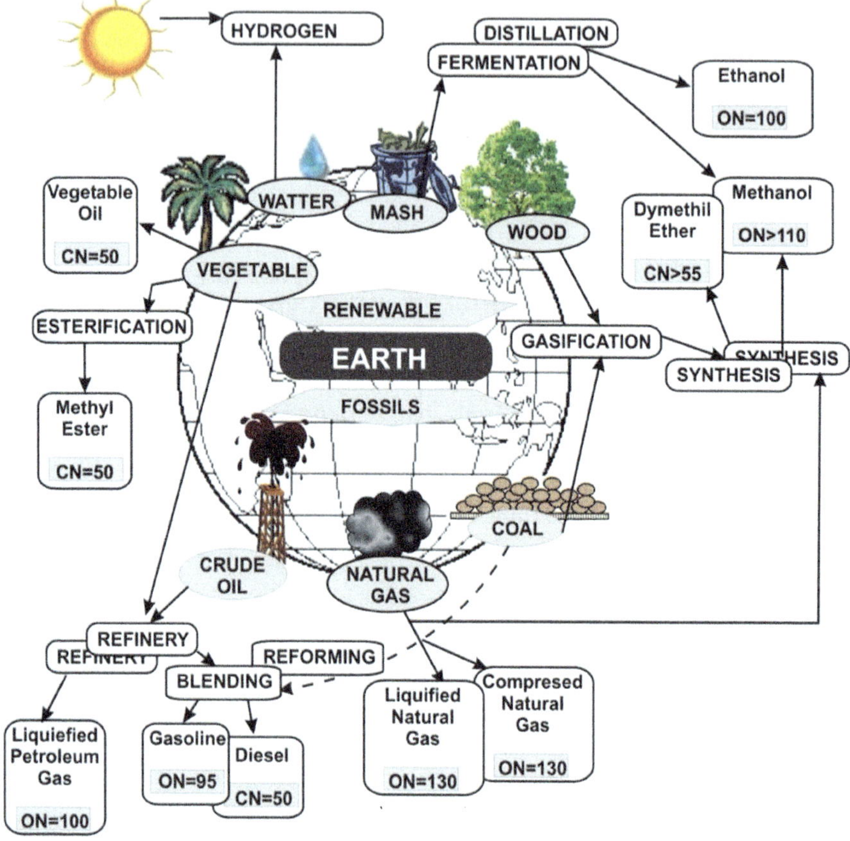

Figure 10.1 Overview of _energy sources and processes_. On the lower side of the picture are the _fossil fuels_, on the upper side the _renewable energy sources_.

When **carbon is present in the molecular structure of an energy source** – fossil or renewable – _carbon_ dioxide _is produced by any form of energy conversion, combustion or any other chemical reaction to_ produce _heat_ or **electrical energy**. The difference between fossil and renewable energy sources is only the _partial recycling_ of the carbon dioxide produced from renewable energy sources.

In the photosynthetic plant nutrition cycle, _carbon dioxide is_ absorbed from the atmosphere.

$$6CO_2 + 6H_2O \xrightarrow{\quad LICHT \quad} C_6H_{12}O_6 + 6O_2$$

Photosynthesis occurs as a complex chain of intermediate reactions, in two main stages [2]:

- In *the light reaction phase*, the chlorophyll in the plant is activated by light absorption, which produces adenosine triphosphate (ATP) and a form of triphosphate pyridine nucleotides (TPN), splitting water to release the hydrogen required for the process.

- In the <u>dark reaction phase</u>, the ATP and TPN components provide the energy for the absorption of carbon dioxide. This produces **carbohydrates**, or various forms of sugar, to feed the plant.

Ethanol production from plants and biowaste

Even though the amount of CO2 absorbed by the plant and stored in the energy carrier is currently lower than the amount of CO2 produced by the processing and combustion of the energy source – from sugar cane cultivation in Brazil to ethanol combustion in the engine of a vehicle, a return rate of around 60% is reported – there is a significant advantage over fossil fuels. The recyclable CO2 content can increase well over 60% if the agricultural machinery and chemical/distillation plants required in the overall process of crop cultivation and processing are converted to renewable energy sources. Another major advantage is the practically unlimited availability of resources, which can cover the entire mobility requirements in terms of quantity, even for the increasing demand in the future.

According to [4] in 2030 the total energy consumption in the world will be 720 exajoules/year [1 exajoules [EJ] = 1.0e+12 MJ] (in 2000 it was 430 exajoules); on the other hand, the agricultural area will grow to 580 million hectares for the production of bioenergy – from sugar beet and rapeseed to beet – and will provide a total energy of 1031 exajoules/year. Occasional opinions according to which, for example, ethanol is of no interest to Germany because the areas in the country that can be planted with energy sources are far from sufficient, have a flaw: According to this view, there should be no Oil-based heat engines are operated with almost 100% dependence on imports.

An increased import of *oils and alcohols from plants, biomass or waste from the wood, paper and cellulose industries* is expected regardless of the shortage of oil and gas reserves. In May 2011, for example, the U.S. Congress passed a law [5] requiring 50% of automobiles produced in the U.S. in 2014 to be fueled from non-petroleum fuels – *alcohols, natural gas, hydrogen, biodiesel.* In 2016, 80% and in 2017 even 95% of automobiles are to be produced for such fuels. In 2017, there were already **50** million ethanol-based vehicles worldwide – 29 million of them in Brazil and 18 million in the USA – that are powered by variable mixtures of ethanol and gasoline (FlexFuel).

Methanol production from industrial carbon dioxide emissions

The use of carbon dioxide to produce alcohol as a fuel is not only through plant regeneration through photosynthesis, but also through recycling of industrial carbon dioxide emissions, as new large-scale projects show.

At 800 million tons per year (2019), **the Federal Republic of Germany has** the highest carbon dioxide emissions compared to all European countries. Of this, **300** million tons come from the energy sector *and* 133 **million** tons from the energy sector.
Emission sources are absolutely comparable: **160** million tons are produced in **industry** and also 160 million tons in **road transport (cars and commercial vehicles).** And hence the approach: road traffic should absorb the carbon dioxide emissions of industry! Due to the efficiencies in the overall chain of *producing hydrogen, absorbing carbon dioxide in filters and then synthesizing methanol, the* balance cannot be complete, but the recycling is nevertheless considerable. This scenario is very realistic and is already being implemented: ThyssenKrupp produces **15** million tons of steel annually in Duisburg, with a carbon dioxide emission of **8** million tons, which means 1% of the total carbon dioxide emission in Germany! In September 2018,

ThyssenKrupp, together with 17 other partners and with the support of the German government, launched a corresponding project.

The *Carbon2Chem* program envisages the capture and implementation of 20
million tons of carbon dioxide per year. The method used for synthesis in *methanol*

Conducive *hydrogen* is **generated electrolytically**. Electrolysis, in turn, requires electrical energy, which will be secured by wind power and photovoltaic systems as part of the project. In contrast to coal-fired or nuclear power plants, these two forms of electricity production are characterized by a strong temporal fluctuation of energy production. However, the correspondingly discontinuous methanol production is not a problem, methanol can be stored in the liquid phase at ambient pressure and temperature.

The *Carbon2Chem* concept can certainly be extended from the factories to the larger emitters – these are the coal- and gas-based power plants, next to which wind power and photovoltaic systems for hydrogen electrolysis can also be installed.

Methanol production from carbon dioxide in garbage

The other useful source of carbon dioxide is garbage: each European produce, on average, 475 [kg] of waste per year. As an example, one of Munich's two large waste incineration plants is supplied with 800,000 tons of coal in one module and the other 2 modules with 650,000 tons of household waste annually. Its combustion generates **900 [MW] of heat** and 411 [MW] **of electrical energy** [3].

In addition, there is a corresponding carbon dioxide emission that could be used for methanol production. There are large waste incineration plants next to every European metropolis, and safeguarding methanol as a fuel, for both gasoline and diesel engines, where there is the greatest density of vehicles, would also provide a logistical advantage.

In this way, electromobility will face healthy competition. Asia is expected to have the largest sales of electric cars – 2 million – but also the largest sales of cars with internal combustion engines – 32.6 million.

A forecast for the share of alternative energy sources [6] in all modes of transport by world region for the year 2050. It is noteworthy that the energetic share of biofuels dominates – whereby the entire transport was considered, on earth, in the air and on water.

As in the case of the addition of natural gas to crude oil, mixed solutions will only be used in bivalent combustion engines due to the infrastructure that has yet to be built. In the current use of CNG (Compressed Natural Gas) and LPG (Liquefied Petroleum Gas), most vehicle manufacturers prefer bivalent solutions - CNG/petrol, LPG/petrol - with separate tanks and injection systems.

10.2 Alternative fuel properties

The use of alternative fuels depends not only on their resources, environmental aspects and the technical and technological basis of their production, but also, to a particular extent, on their characteristics on board a vehicle. The most important relationships between fuel properties and engine or vehicle characteristics are presented as follows:

Molecular Structure of Fuel (CmHnOp):

It directly affects the *structure* and *concentrations of exhaust gas components* as a result of the mass fractions of carbon and hydrogen in the combustion reaction in any plant – vehicle engine, heating system, industrial firing.

For example, the combustion of *carbon (C1H0O0) results* in the maximum CO_2 concentration per kilogram of fuel; on the other hand, the combustion of hydrogen (C0H1O0) does not produce carbon dioxide, but only water.

Density of fuel:

The volume, but also the mass of the overall fuel tank system, depends on this.

Gasoline, diesel, methanol, ethanol and the *oil esters* have a largely similar <u>density</u> in terms of ambient pressure and temperature; the existing tank systems are suitable for this without significant modifications.

Due to its molecular structure, hydrogen poses a considerable problem for use in mobile applications in terms of storage capacity.

By way of comparison, under the same pressure and temperature conditions, [1] 14.5 times less hydrogen mass than air mass could be stored in the same tank volume! According to the equations of state in [1] the stored mass in a given reservoir can only be increased if the *pressure is increased* on the one hand and the temperature is lowered *on the other*. As an example of these concepts, cryogenic hydrogen storage is considered at *-253* [°C] / 0.1 *[*MPa]; even if at this considerable temperature the hydrogen becomes liquid, its density is just one-tenth of the gasoline density!

Viscosity of the fuel:

It essentially influences the parameters in the fuel metering system – through the lubrication properties, but also on the combustion process. For example, injection pressures in the area of diesel fuel using conventional plungers – such as in common rail or pump-nozzle systems – with gasoline are not achievable at all due to the lubricating properties. Again, show according to this criterion, *petrol, methanol and ethanol* have similar properties, which speaks in favor of the same fuel injection technology.

The viscosity of oils is twenty times higher than that of diesel fuel, which allows them to be used directly only in rare cases. Even <u>transesterification</u> – the shortening of the molecules – is associated with twice the viscosity of diesel fuel. Regardless of the lubrication in the injection system itself, viscosity – as an expression of long, branched molecules – impairs combustion, through local oxygen deprivation. This explains the coking phenomena when using pure vegetable oils, for example in piston engines.

Calorific value of the fuel:

The *heat* that can be obtained as a result of the exothermic chemical reaction of a kilogram of fuel depends on its <u>elementary structure</u>, i.e. on the mass fractions of *carbon, hydrogen and oxygen* in its molecule. The higher the calorific value, the less fuel mass for a comparable energy *(in the case of vehicles as <u>a performance profile</u> over a period of time)*. With the same tank capacity (as fuel mass) and the same performance profile, the calorific value of the fuel results in the range.

An example, again from the vehicle sector: With the same fuel mass, the range of the same vehicle with the same combustion engine, which corresponds to the same power requirement, would be about three times greater with hydrogen than if gasoline were used.

Again, the transition from gasoline to methanol would reduce the range by half with the same tank capacity. The latter case largely corresponds to real conditions, due to the comparable density of gasoline and methanol under the same storage conditions (ambient state).

Otherwise, a comparable tank volume – due to the space available in the vehicle – must be assumed rather than a comparable stored mass of fuel. From this point of view, the advantage of hydrogen in terms of range changes into a disadvantage: the calorific value compared to gasoline is about one-tenth of the density (in the liquid phase).

With the same tank volume (in the case of hydrogen with the specified storage pressure or temperature values), the range of a vehicle when switching from gasoline to hydrogen remains about a third of the value when running on gasoline.

Stoichiometric air requirement of fuel:

The air requirement, like the calorific value, depends on the elementary structure of the fuel. Hydrogen requires the largest air mass for a stoichiometric (chemically exact) reaction. Alcohols already contain a proportion of oxygen, which is why they use less oxygen from the ambient air in combustion than petrol. In the case of combustion in stationary plants, the fuel mass flow of a specific fuel can be adjusted by means of the respective dosing system – the air mass flow is then proportionally adjusted to the air demand by a corresponding flow system. In piston engines, the cylinder space is limited as the sum of the displacement and the combustion chamber of all n cylinders (V=n(Vh+Vb)), i.e. it can only include a certain mass from the environment.

In the case of a recharge, the same charge pressure p is assumed. For the same amount of air in an engine, the stoichiometric air demand changes the amount of fuel that can be supplied under chemically exact conditions in the opposite way: the greater the air demand, the less fuel is supplied with the same amount of air. When switching from gasoline to methanol in the same engine, the amount of methanol supplied increases by 2.2 times compared to the amount of gasoline, which must be taken into account by the parameters of the injection system (fuel pressure, opening time of the nozzles, flow cross-sections). On the other hand, when switching from gasoline to hydrogen, the amount of fuel is reduced by a factor of 0.42.

Calorific value of the fuel/air mixture:

The reduction in the quantity supplied of a fuel with a high calorific value as a result of the stoichiometric air demand for the same amount of air [*kg of fuel per kg of air] impairs* its energy advantage: although the calorific value of hydrogen in itself - exceeds the calorific values of all other fuels - which theoretically leads to the expectation of more heat supplied and thus more displacement-related power - the high air demand significantly reduces this advantage: The calorific value of the hydrogen-air mixture is lower than that of a hydrogen-air mixture.

Gasoline-air mixture. In this comparison, the efficiency of combustion was kept the same. In the real process, more efficient combustion is possible due to the evaporation properties of the hydrogen and its molecular size, which compensates for or partially reverses the disadvantage compared to gasoline-air mixtures. An average mixture calorific value appears to be largely realistic for all the fuel-air mixtures listed.

This means that the switch from a conventional to an alternative fuel in a combustion plant does not lead to a spectacular change in the specific work [kJ/kg mixture].

Octane number, cetane number of fuel:

The knock resistance in gasoline engines – expressed by the octane number – and the ignitability in diesel engines – expressed by the cetane number – is characterized by noticeable differences due to the molecular structure, but also the evaporation behavior of the fuels under consideration. From this point of view, natural gas and alcohols appear to be more advantageous than gasoline: raising the knock limit at a higher octane rating allows the compression ratio to be increased in gasoline-based piston engines, resulting in an improvement in thermal efficiency. This reduces specific fuel consumption. In the diesel process, the ignitability of the dimethyl ether – caused by its evaporation properties, but also by the oxygen in the molecular structure – is remarkable. On the other hand, pure, non-esterified oils are less susceptible to ignition due to long, branched molecules; the result of transesterification with respect to this property is convincing in this regard.

Evaporation enthalpy of fuel:

Fuel evaporation is a major criterion of mixture formation quality and therefore combustion; the smaller the fuel droplets and the more they change from liquid to gaseous phase, the more favorable the conditions for efficient and complete combustion. The enthalpy of evaporation is extracted from the surrounding air in a usually adiabatic mixture formation process, which lowers its internal energy and thus the temperature. There is only a restricted potential in cold air, which is why fuels with a high required evaporation enthalpy – especially methanol – have relatively poor cold-start properties in the event of mixture formation in the intake manifold. The internal mixture formation by means of direct fuel injection changes the evaporation process in a remarkable way: the injection in the cylinder takes place on much warmer air – partly through compression, partly through thermal radiation from the combustion chamber walls. The evaporation process intensifies between the injected, evaporable fuel and the air that has already been heated. The higher heat flux not only promotes fuel evaporation, but also achieves a more significant reduction in air temperature, which in turn accommodates an increase in the compression ratio, for example, in piston engines.

Due to their high enthalpy of evaporation, combined with the relatively fast evaporation, ethanol and methanol appear to be ideal fuels for injection in a combustion plant.

10.3 Methanol and ethanol: production, properties, storage, use

Nikolaus August Otto used ethanol in his engine prototypes as early as 1860, Henry Ford used *bioethanol in production vehicles between 1908 and 1927 and described it as the fuel of the future.*

Ethanol and methanol are obtained from two groups of raw materials:

Starch and sugar – from plants or plant residues. In Latin America, especially in Brazil, sugar *cane molasses is used for this purpose, in* North America *corn, in Europe* sugar beet *and partly wheat, and in Asia* cassava (cassava).

A newly explored, globally available raw material base for ethanol consists *of algae, cellulose – residues from the paper or wood processing industry,* plant *waste and plants that are unsuitable for human consumption.*

The availability of such raw materials and their impact on the human and natural environment when systematically used as an energy source for mobility can be illustrated with the following facts and examples:

Sugar cane has been cultivated in Brazil since 1532. Ethanol from sugar cane was used there as a fuel for automobiles between 1925 and

1935. Since 1975, after the 1st global oil crisis, the Brazilian government has consistently pursued the national alcohol program ProAlcool to replace fossil fuels with alcohol. The first mass-produced car powered by 100% ethanol since the introduction of the ProAlcool program was the Fiat 147 (1979); ten years later, 4 million vehicles were running on 100% ethanol in Brazil. The reversal of this trend towards greater dependence on oil in the following years was primarily due to cross-border economic policy causes. However, this situation was overcome relatively quickly. From 2003

The Brazilian VW Gol 1.6 Total Flex was launched, a car for variable mixtures (0-100%) of gasoline and ethanol (Flex Fuel). Seven years later, *Chevrolet, Fiat, Ford, Peugeot, Renault, Volkswagen, Honda, Mitsubishi, Toyota, Citroen, Nissan and Kia* were on the Brazilian market with flex fuel cars, accounting for 94% of all new registrations.

There are currently 29 million Flex Fuel vehicles on the road in Brazil (2017). This intensive use of ethanol from sugar cane could theoretically lead to the problem of raw material availability. However, Brazil has **355** million hectares of arable land, of which **72** million hectares are currently ploughed. Sugar cane is planted on only 2% of arable land, 55% of which is used for ethanol production. Brazilian scientists believe that cane sugar cultivation can be increased 30 times without harming the environment and also without endangering food production [3]. Productivity is up to *8,000 litres of ethanol per hectare* (2008) at a price of 22 *US cents/litre*. Ten times more energy is produced – in the form of ethanol fuel – than the energy used in the entire process between sugar cane cultivation and the production of the corresponding amount of ethanol.

99.7% of the sugar cane plantations are located on plains in the southeastern region of Sao Paolo, i.e. at least *2,000 [km]* from Amazonian tropical forest, where the climate is rather unsuitable for sugar cane.

In the U.S., ethanol is mainly produced from grain and corn. This requires 10 million hectares, which is 3.7% of arable land. The productivity is up to *4000 liters of ethanol per hectare* (2008), which is half that of production from sugar cane in Brazil. The ethanol production is only 1.3 to 1.6 – low compared to the value of 10 for sugar cane. At *35 US cents/litre,* the production price is higher than the use of sugar cane *(22 cents). Ford, Chrysler* and *GM* build flex fuel powertrains in their entire range of vehicles – from sedans and SUVs to off-road vehicles. There are currently *10 million Flex Fuel vehicles on the road in the United States.* A current U.S. government program for the next few years envisages the increased production of *cellulosic ethanol from agricultural residues, from residues from the paper industry* as well as from *household waste.*

In addition to *sugar cane, grain, maize, sugar beet and cassava, algae* represent an important raw material potential for the production of alcohol. Algae are aquatic creatures that feed on the basis of photosynthesis. The yield per area – albeit when cultivated in algae reactors – is significantly higher than for the production of biomass in agriculture – 15 times higher than rapeseed and 10 times higher than maize.

Research is currently very active in this area – two companies are significant in this context: *Boeing and Exxon.*

Methanol and ethanol can be produced by two methods:

- Distillation of fermented biomass

- Synthesis, gasification and reaction by cyanobacteria and enzymes

The natural formation of alcohol in the fermentation of sugary fruits was already noted by our ancestors, as mentioned in ancient Egyptian and Mesopotamian writings, but also in the Bible. The production of alcohol from biomass is similar to that used for the production of fruit *brandy, rum, whisky, vodka or* sake – representing all continents – from fruit or vegetables. In Japan, sake was extracted from fermented rice as early as the 3rd century BC. In the 10th century, the distillation of wine from lychee and plums for the production of high-proof brandy was widespread in Anatolia (Asia Minor). The overproduction of grain in the mid-18th century led to a large-scale production of gin in England.

The simplest form of *distillation* is the boiling of fruit, which is cooked within a few weeks of free storage, followed by condensation of the resulting steam by cooling the steam pipe externally – for example, with a stream of cold water – and supplying the resulting liquid alcohol to a vessel: first methanol, then ethanol. This simple illustration is only intended to emphasize that such a technology is easy and well controllable and that it can be used anywhere in the world in large-scale plants or decentralized. Industrially, the sugary mash from the fermented raw material, which already has an alcohol content of around 10%, is brought to a concentration of more than 99% by *distillation/rectification*.

A particularly interesting alternative is the production of ethanol from *waste containing hydrocarbons,* such as old *tyres* or *plastic containers*, as well as *biowaste*. The process developed by Coskata is shown in Figure 10.2.

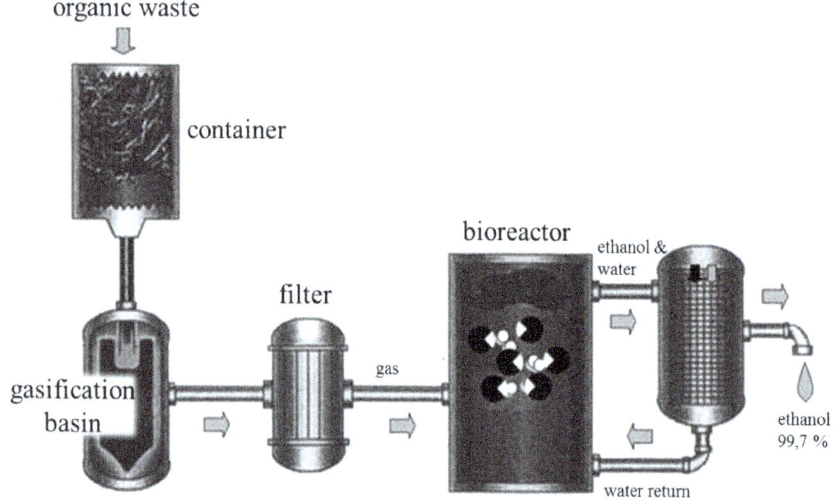

Figure 10.2   Process for ethanol production from waste - schematic

The hydrocarbon structures in the waste are converted into a synthesis gas by cracking. The chemical energy in the proportions included carbon dioxide and Hydrogen is then used by microorganisms in a bioreactor to produce ethanol.

The microorganisms show an increased tolerance to impurities that would inhibit classical chemical transformation. According to _General Motors_ as the promoter, the production cost of ethanol using this process remains below one US dollar per gallon – and thus about half the production cost of gasoline. One gallon of water is required to produce one gallon of ethanol using this process, which is one-third of the amount of water required to produce conventional biofuels.

In 2011, Coskata planned to produce 50 to 100 million gallons of ethanol per year in its first large-scale commercial plant. In 2020, 18% of the oil demand in the USA could be replaced by ethanol produced by this process.

The efficiency chain between the energy source and the driven wheel of a vehicle would lead to an 84% reduction in CO2 emissions through this form of recycling compared to the use of petrol.

Recently, as described in chapter 10.1, the synthesis of methanol from carbon dioxide produced by industry and from climate-neutral hydrogen produced in large quantities has been realized.

In 1998, around 216,000 Flex Fuel vehicles were produced in the USA, in 2012 more than 10 times that number (2.47 million), and the current figure of 15.11 million proves the acceptance of this concept by customers beyond the respective government programs.

The use of methanol in gasoline and diesel engines is also promising.

Gasoline engines with methanol manifold injection are used on a large scale in China: in the metropolis of Xi'an (12 million inhabitants/2017), 80% of the 10,000 taxis with gasoline engines run on 100% methanol.

Large diesel engines using the four-stroke process, with 100% methanol direct injection, were developed and put into series production by Wärtsila for use in ships. B&W/MAN has also implemented diesel engines with methanol direct injection for marine use. In this case, however, we are talking about _two-stroke engines._

Both *the four-stroke engines from Wärtsila and* the two-stroke engines from B&W/MAN no longer use the classic diesel compression-ignition, but the pilot direct injection of a small amount of diesel fuel, also because methanol has a cetane number of only 3 (diesel: 50 to 60).

The Wärtsila *four-stroke pilot-injected methanol diesel engines can be found on the Stena Germanica (2015), while* the pilot-injected two-stroke methanol diesel engines on the Lindager ship provide 10,320 [kW] of power.

The application of mixtures of gasoline and alcohols from plants and biomass for direct injection in piston engines offers considerable potential with relatively little effort for future developments. The advantages of such a concept appear to be compelling:

- Annual *renewable energy sources, recirculation of carbon dioxide in the plant cycle*, reduction of specific fuel consumption through direct injection, use of existing infrastructure through variable fuel proportions depending on availability.

In the longer term, the use of alcohols appears to be a major source of energy, even its production cycle is also used as an energy source. It is just as environmentally friendly as the use of electrolytically produced hydrogen.

The main energy source and the process chain are similar, only the energy-carrying component is different:

- ***Alcohols (methanol, ethanol):*** The energy of the **sun's radiation** is used in one way for propulsion by means of carbon dioxide, which is formed during combustion and split again in the plant – as a natural reactor. *The carbon dioxide in nature is used as a carrier for energy conversion,*

- ***Hydrogen:*** In the other way, the energy of **solar radiation** is used for propulsion resulting in water, which is formed during combustion and electrolytically split again in an industrial plant. In this case, the water in nature is used as an alternative to carbon dioxide as a carrier for energy conversion.

The only significant difference between the two circuits is the system for splitting the respective molecule – CO2 or H2O – into an energy carrier.

The carbon dioxide splitting plant is provided by nature itself.

10.4 Hydrogen: production, properties, storage, use

Hydrogen is the subject of most ideal scenarios for the propulsion systems of the future:

Its production is theoretically possible on the basis of solar energy or by electrolysis from water, without carbon dioxide and no harmful by-products.

Its use – either by combustion in a heat engine or by proton exchange and thus electricity generation in a fuel cell – leads back to the original water – as *long as no NOX is produced by the nitrogen in the air involved due to dissociation at high combustion temperatures in heat engines.*

The splitting of water in a fixed place, in an efficiently designed plant, and its restoration in a propulsion system appears to be the ideal form of energy transfer by storing an intermediate product as an energy carrier, in this case hydrogen.

The problem, however, is the storage itself, which requires either high pressure or extremely low temperature for a storage mass due to the hydrogen-gas constant R [kJ/kg hydrogen*K] of the – the largest of all elements.

Air has a specific gas constant of 287.04 [J/kg. K] – 14.4 times less than hydrogen, so there would be 95 grams of air in the same tank under the same pressure and temperature conditions. Again, in an 80-litre petrol tank, for example in a car with a petrol engine, which is to be operated with either hydrogen or petrol – with a petrol density of 0.75 [kg/liter], there would be room for a mass of 60 [kg] of petrol.

This thermodynamic relationship remains independent of technical progress in storage. In addition, the production of hydrogen is practically not carried out according to the ideal, clean scenario *of water splitting – hydrogen storage: Currently*, only less than 1% of hydrogen is produced in this form worldwide – electrolytically, using *solar, water and wind energy*. The vast majority – over 99% of the 500 billion standard cubic meters produced annually – is derived from fossil fuels; this produces carbon dioxide or carbon [2]. The hydrogen produced in this form has been used for decades in industry for the production of *fertilizers, paints, solvents and plastics, as* well as in the *petroleum industry to improve fuel structure.*

One of the main problems with the use of hydrogen in plants is its storage on board, which is only possible at one-tenth of the gasoline density, even in the liquid phase temperature of *-253* [*°C*]). The liquefaction itself requires a third of the energy stored in the respective amount of hydrogen. The alternative option - *high pressure instead of low temperature* - is also used, at pressure values between *35 and 90* [*MPa*], where the hydrogen remains in the gaseous phase. It should be noted, however, that the difference to the ambient pressure creates an outflow potential: the hydrogen molecule is the smallest of all elements and can therefore easily penetrate most material structures. This requires a multi-layered, complex design of a hydrogen tank, which also has to have the appropriate strength at the high pressure, which is currently considered solved. Due to the high air/fuel ratio, the volume-related mixture calorific value is lower than that of petrol/air mixtures, which leads to low specific work for the same displacement.
In connection with the properties described, storage is usually carried out in liquid form, in cryogenic tanks, at *-253* [*°C*], for tank volumes that are considerably larger than conventional gasoline tanks.

Two tank layers are separated by a gas layer, the gaseous hydrogen is accumulated, cryogenic storage requires a cooling circuit with an appropriate heat exchanger.

Stationary storage in the filling station, the transport itself, does not cause any problems due to the long tradition of hydrogen use in industry.

On the other hand, due to the very low density, that is, the very large volume, even at a low mass, which requires injection of four times the volume of hydrogen compared to gasoline, for example, in the case of direct injection in piston engines, either a remarkable increase in injection pressure or a significant increase in injection time. The room for maneuver is relatively limited in both directions.

The contact of the injector with the combustion chamber requires, in turn, a much more effective thermal insulation than in the case of manifold injection. The shell of the nozzle body comes into contact with a temperature of around *200 [°C] instead of* 15-20 [°C] during manifold injection. Maintaining a hydrogen temperature of *-253 [°C]* until it comes out of the injector to protect the liquid phase reinforces the heat transfer and creates insulation problems.

For example, there are currently around 600 hydrogen-powered cars in use worldwide. The absolute pioneering work, from the exploration of the potentials to the standard use in automobiles, is due to BMW. The first hydrogen car was already running at BMW in 1979, with a 4-cylinder engine that reached an output of *60 [kW]*. Parallel to the development of hydrogen cars, BMW worked on the production of hydrogen using solar energy and water electrolysis as part of partnerships such as *Solar-Wasserstoff-Bayern* GmbH. BMW used a compact fuel cell in the vehicle, parallel to the drive motor, for power generation, which had an electrical output of *5 [kW] at a voltage of* 42 *[V]*.

10.5 Dimethyl ether: production, properties, storage, use

Dimethyl ether is an alternative to diesel fuel. It can be produced from natural gas or coal, but its production from wood waste seems promising, although the production process *is similar to* that of methanol. The steps are gasification and then synthesis. The availability of the energy source and

The CO2 cycle, similar to the use of alcohol, is supplemented by further advantages when the dimethyl ether is used in diesel processes:

- The high oxygen content of about 35 % (mass) can be expected to improve carbon combustion and thus reduce soot and particle emissions, in line with the combustion behavior of oil esters.

- Its low compression-ignition temperature of 235 [°C], also expressed as a high cetane number, has the advantage of a more favorable combustion process and thus a higher thermal efficiency than when using diesel fuel.

According to the data, dimethyl ether in the liquid phase reaches a density that is about 15% lower than that of diesel fuel. Such a value is quite acceptable in comparison with the other alternatives listed. The viscosity is far below the values for diesel fuel and thus causes similar problems as vegetable oils and oil esters, only from the other side: The lubrication of the plungers in a fuel pump is practically impossible with this fuel – this requires compensatory measures. Due to the similar ratios of *carbon, hydrogen and oxygen in* the molecule as in ethanol, the calorific value, *air demand and mixture calorific value* of dimethyl ethers and ethanol are almost identical. The calorific value mixture is therefore slightly lower than that of diesel fuel/air mixture, but can be compensated by better evaporation and combustion in terms of the specific work that can be achieved.

Liquid dimethyl ether can be stored at *20 [°C] under a relatively low pressure of* 0.5 *[MPa]*. Liquefied gases are characterized by a relatively high

On the other hand, if the vapor pressure falls below the pressure, there is a risk of gas bubble formation, which jeopardizes the functioning of the injection system. Therefore, a pump is required in the tank of such a system that always keeps the pressure above the steam pressure. In addition, leaks occur in the injection system due to their function, which reach the steam pressure in the return flow. This means that the fuel changes phase from liquid to gas after the leakage site. The gas must be collected in a separate tank (purge tank) and pumped back above the steam pressure by means of a pump and then transferred to the main tank.

Tests with dimethyl ether in diesel engines with *1* or *2* [*dm³*] per cylinder, but also with passenger car diesel engines, show excellent engine results: Die ULEV (*Ultra Low Emission Vehicle*) emission standard can be achieved for the larger cylindrical can be achieved without exhaust gas aftertreatment or, in the case of passenger car engines, with a simple oxidation catalytic converter [2]. The thermal efficiencies do not differ when operating with dimethyl ethers or diesel fuel, but the noise emission can can be drastically reduced with the use of dimethyl ethers. In addition to alcohols, hydrogen and oil esters, dimethyl ether represents a promising alternative as an environmentally friendly and regenerative energy source for automotive drives is a promising alternative as an environmentally friendly and regenerative energy source for automotive propulsion.

10.6 Biofuels: resources, processes, uses

Synfuel, or designer fuel, is increasingly becoming an expression of a new trend in the research and development of energy carriers whose molecular structure can be specifically constructed. Long and extensive experience in process engineering – from *refineries and distilleries to gas synthesis, transesterification, pyrolysis, electrolysis or other thermochemical or electrochemical processes* – lead to new, effective combination possibilities that pave the way to the controlled construction of molecular structures.

The production is carried out according to the Carbo-V/Fischer-Tropsch process, by hydrogenation process or by pyrolysis process:

-   *– according to the Carbo-V/Fischer-Tropsch (BtL) process*, in a first stage, the biomass is gasified in a reactor with the addition of heat and at a given pressure with the participation of oxygen. The process consists of three stages:

-   Low-temperature gasification in which the biomass with a water content of 15-20% is carbonized by partial oxidation with air or oxygen at temperatures in the range of 400-500 [$°C$], producing a tar-containing gas and solid carbon (biocoke).

- <u>sub stoichiometric oxidation of the tar-containing gas</u> above the melting point in a combustion chamber.

- <u>Injection of the biocoke ground into dust into</u> the hot gasifier, whereby the synthesis raw gas is produced in the gasification reactor within an endothermic reaction. The synthesis gas consists mainly of CO2, CO, H2.

In the subsequent Fischer-Tropsch synthesis, the reactive components of the synthesis gas (CO, H2) are formed in a catalyst - based on Iron, Magnesiumoxid-, Thoriumoxid- or Cobalt - Hydrocarbon chains.

An engine with a given displacement must therefore increase the amount of fuel injected for reasons of chemical balance when the air requirement is reduced. In such a case, the increase in the amount of fuel injected has nothing to do with an increase in fuel consumption as a result of poorer combustion.

10.7 Fatty acid methyl ester (FAME)

FAME (*fatty acid methyl ester*) consists of compounds of a *fatty acid* and *methanol*. To produce *fatty acid methyl ester* (FAME), the trivalent alcohol **glycerol** is exchanged for **methanol**. Glycerol and FAME are produced as reaction products, whereby, when natural fats are used, a mixture of different fatty acid methyl esters is always produced from *natural triglycerides (basically several different fatty acids)*. A mixture of FAME, which consists of vegetable fats (rapeseed oil) or animal fats (from lard) and methanol, is also known as **"biodiesel"**. FAME made from vegetable fats are liquid at room temperature. The storage of FAME should take place with as little oxygen as possible, since FAME reacts with oxygen from the air duc to the double bonds in the long-chain unsaturated fatty acid methyl esters [7] and can become resinous by bridging the individual molecules.

10.8 Hydrogenated Vegetable Oils (HVO)

Hydrogenated vegetable oils are produced in existing refineries together with other fats and mineral oil components as well as in our own vegetable oil plants.

During hydrogenation in mineral oil refineries, vegetable oils such as rapeseed oil are added to the vacuum gas oil produced during the processing of the mineral crude oil in proportions of up to 30 percent.

In the subsequent hydrotreating, these vegetable oils are then chemically modified together with the mineral oil fraction by removing the so-called heteroatoms such as *sulphur, oxygen and nitrogen with the* incorporation of **hydrogen**. In addition to the hydrocarbons produced from the vegetable oils, hydrogen sulfide (H2S), water (H2O) and ammonia (NH3) are produced as by-products. In a subsequent step (hydrocracking), the hydrocarbons are again split into smaller chains (cracking) with hydrogen incorporation, producing methane (CH4), propane (C3H8) and water as by-products.

10.9 Synthetic fuels: resources, processes, use

Synfuel, or designer fuel, is increasingly becoming an expression of a new trend in the research and development of energy carriers whose molecular structure can be specifically constructed. Long and extensive experience in process engineering – *from refineries and distilleries to gas synthesis, transesterification, pyrolysis, electrolysis or other thermochemical or electrochemical processes* – lead to new, effective combination possibilities that pave the way to the controlled construction of molecular structures.

The criteria for the design and use of an ideal synthetic fuel in *fuel cells* are generally similar to those for use in *internal combustion engines*. In this case, however, what counts most is the rapid and not particularly complex release of the hydrogen in front of the surface where the proton exchange with the oxygen takes place.

A molecular structure of the CmHnOp type does not appear to be optimal in this case. Rather, the hydrogen should be enclosed in a matrix structure in a liquid phase, whereby at the reaction site along the proton exchange surface, the matrix elements should release the hydrogen rapidly under given pressure and temperature conditions, but behave chemically neutrally. An analogy with the process of transporting carbohydrates or glucose from blood to the muscles by means of insulin in the bodies of living beings can lead to new methods.

Thermodynamical cycle processes and suitable machines were analyzed, which demonstrate a high thermal efficiency with excellent displacement-related performance for stationary use as power generators in hybrid systems.

An example, which implies other interesting energy sources, is particularly interesting:

A stationary Stirling engine requires a heat source, which runs off by external combustion, at constant thermodynamic and fluid mechanical conditions. The process can be similar in a turbomachine (*gas turbine*) in stationary operation if the combustion chamber is replaced with a heat exchanger in conjunction with external combustion.

From this perspective, a potential is opened up which, on the one hand, has a high energy density during storage, and on the other hand, high combustion temperatures: the combustion of metal powders. The combustion of aluminum powder at very high temperatures is well known from welding technology. The ignition of magnesium in the air above *500* [°*C*] has long prevented the casting of magnesium parts, in this direction its use – as magnesium powder – is turning from a disadvantage into an advantage. Iron powder reacts similarly with the oxygen in the air. Mixtures of such solid fuels, such solid fuels, in particular the presence of proportions of aluminum powder, are used in modern rocketry due to their excellent energy density. Mixtures of metal powders in defined proportions are also to be considered as <u>synthetic fuels</u>. However, they are less complex than liquid molecules. From this perspective, too, alternative heat engines with external combustion as power generators on board a hybrid vehicle appear to be realistic and promising.

References for Chapter 10

[1] Stan, C.: Thermodynamics for Mechanical Engineering and Vehicle Construction, 4th edition, Springer, 2020, ISBN 978-3-662-61789-2

[2] Stan, C.: Alternative Propulsion for automobiles, Springer, 2020, ISBN 978-3-662-61757-1

[3] Stan, C.: Energy versus carbon dioxide, Springer, 2021, ISBN 978-3-662-62705-1

[4] FAO Prognosen (*Food and Agriculture Organization of the United Nations*), https://data.apps.fao.org/map/catalog/static/search?keyword=ethanol

[5] the Open Fuel Standard Act (OFS) https://www.govinfo.gov/content/pkg/BILLS-112hr1687ih/pdf/BILLS-112hr1687ih.pdf

[6] Study by the International Energy Agency (IEA) https://ourworldindata.org/agricultural-production

[7] https://de.wikipedia.org/wiki/Fetts%C3%A4uremethylester

Index

Printed by Printforce, the Netherlands